Materials and Processes for Photocatalytic and (Photo)Electrocatalytic Removal of Bio-Refractory Pollutants and Emerging Contaminants from Waters

Materials and Processes for Photocatalytic and (Photo)Electrocatalytic Removal of Bio-Refractory Pollutants and Emerging Contaminants from Waters

Editor

Annalisa Vacca

MDPI • Basel • Beijing • Wuhan • Barcelona • Belgrade • Manchester • Tokyo • Cluj • Tianjin

Editor
Annalisa Vacca
Università di Cagliari
Italy

Editorial Office
MDPI
St. Alban-Anlage 66
4052 Basel, Switzerland

This is a reprint of articles from the Special Issue published online in the open access journal *Catalysts* (ISSN 2073-4344) (available at: https://www.mdpi.com/journal/catalysts/special_issues/Electrocatalytic_(Photo)Electrocatalytic_Removal_Pollutants).

For citation purposes, cite each article independently as indicated on the article page online and as indicated below:

LastName, A.A.; LastName, B.B.; LastName, C.C. Article Title. *Journal Name* **Year**, *Volume Number*, Page Range.

ISBN 978-3-0365-3559-3 (Hbk)
ISBN 978-3-0365-3560-9 (PDF)

© 2022 by the authors. Articles in this book are Open Access and distributed under the Creative Commons Attribution (CC BY) license, which allows users to download, copy and build upon published articles, as long as the author and publisher are properly credited, which ensures maximum dissemination and a wider impact of our publications.

The book as a whole is distributed by MDPI under the terms and conditions of the Creative Commons license CC BY-NC-ND.

Contents

About the Editor vii

Annalisa Vacca
Materials and Processes for Photocatalytic and (Photo)Electrocatalytic Removal of Bio-Refractory Pollutants and Emerging Contaminants from Waters
Reprinted from: *Catalysts* **2021**, *11*, 666, doi:10.3390/catal11060666 1

Pedro Martins, Sandro Kappert, Hoai Nga Le, Victor Sebastian, Klaus Kühn, Madalena Alves, Luciana Pereira, Gianaurelio Cuniberti, Manuel Melle-Franco and Senentxu Lanceros-Méndez
Enhanced Photocatalytic Activity of Au/TiO$_2$ Nanoparticles against Ciprofloxacin
Reprinted from: *Catalysts* **2020**, *10*, 234, doi:10.3390/catal10020234 3

Samer Khalaf, Jawad H. Shoqeir, Filomena Lelario, Sabino A. Bufo, Rafik Karaman and Laura Scrano
TiO$_2$ and Active Coated Glass Photodegradation of Ibuprofen
Reprinted from: *Catalysts* **2020**, *10*, 560, doi:10.3390/catal10050560 23

Charlie M. Kgoetlana, Soraya P. Malinga and Langelihle N. Dlamini
Photocatalytic Degradation of Chlorpyrifos with Mn-WO$_3$/SnS$_2$ Heterostructure
Reprinted from: *Catalysts* **2020**, *10*, 699, doi:10.3390/catal10060699 41

Cécile Machut, Nicolas Kania, Bastien Léger, Frédéric Wyrwalski, Sébastien Noël, Ahmed Addad, Eric Monflier and Anne Ponchel
Fast Microwave Synthesis of Gold-Doped TiO$_2$ Assisted by Modified Cyclodextrins for Photocatalytic Degradation of Dye and Hydrogen Production
Reprinted from: *Catalysts* **2020**, *10*, 801, doi:10.3390/catal10070801 63

Laura Mais, Simonetta Palmas, Michele Mascia and Annalisa Vacca
Effect of Potential and Chlorides on Photoelectrochemical Removal of Diethyl Phthalate from Water
Reprinted from: *Catalysts* **2021**, *11*, 882, doi:10.3390/catal11080882 79

Adeem Ghaffar Rana and Mirjana Minceva
Analysis of Photocatalytic Degradation of Phenol with Exfoliated Graphitic Carbon Nitride and Light-Emitting Diodes Using Response Surface Methodology
Reprinted from: *Catalysts* **2021**, *11*, 898, doi:10.3390/catal11080898 91

Alicia L. Garcia-Costa, Andre Savall, Juan A. Zazo, Jose A. Casas and Karine Groenen Serrano
On the Role of the Cathode for the Electro-Oxidation of Perfluorooctanoic Acid
Reprinted from: *Catalysts* **2020**, *10*, 902, doi:10.3390/catal10080902 107

Filomena Lelario, Giuliana Bianco, Sabino Aurelio Bufo and Laura Scrano
Simulated Ageing of Crude Oil and Advanced Oxidation Processes for Water Remediation since Crude Oil Pollution
Reprinted from: *Catalysts* **2021**, *11*, 954, doi:10.3390/catal11080954 121

Pornnaphat Wichannananon, Thawanrat Kobkeatthawin and Siwaporn Meejoo Smith
Visible Light Responsive Strontium Carbonate Catalyst Derived from Solvothermal Synthesis
Reprinted from: *Catalysts* **2020**, *10*, 1069, doi:10.3390/catal10091069 143

About the Editor

Annalisa Vacca, a graduate in Chemical Engineering and PhD in Industrial Engineering, was granted the position of full professor in the field of "Fundamentals of chemical technology" in 2022. She carries out didactic and scientific research in the Department of Mechanical, Chemical and Materials Engineering at the University of Cagliari. Her research activities are focused on the field of electrochemical engineering applied to the study of processes for environmental remediation and energy conversion. In particular, her studies cover key aspects such as the catalytic activity of electrode materials and the identification of reaction mechanisms, as well as practical aspects such as the design and characterization of electrochemical reactors.

Editorial

Materials and Processes for Photocatalytic and (Photo)Electrocatalytic Removal of Bio-Refractory Pollutants and Emerging Contaminants from Waters

Annalisa Vacca

Dipartimento di Ingegneria Meccanica, Chimica e dei Materiali, Università di Cagliari, Piazza D'armi, 09123 Cagliari, Italy; annalisa.vacca@dimcm.unica.it

Citation: Vacca, A. Materials and Processes for Photocatalytic and (Photo)Electrocatalytic Removal of Bio-Refractory Pollutants and Emerging Contaminants from Waters. *Catalysts* **2021**, *11*, 666. https://doi.org/10.3390/catal11060666

Received: 28 April 2021
Accepted: 21 May 2021
Published: 24 May 2021

Publisher's Note: MDPI stays neutral with regard to jurisdictional claims in published maps and institutional affiliations.

Copyright: © 2021 by the author. Licensee MDPI, Basel, Switzerland. This article is an open access article distributed under the terms and conditions of the Creative Commons Attribution (CC BY) license (https:// creativecommons.org/licenses/by/ 4.0/).

This volume is focused on materials and processes for the electro- and photoelectrochemical removal of biorefractory pollutants and emerging contaminants from waters to show the importance of electrochemistry and photoelectrochemistry in offering solutions to current environmental problems. In addition, we highlight their interdisciplinarity and emphasize the fundamental and applied aspects of these methods.

The research for innovative methods for removing pollutants from water has grown along with the detection of new contaminants in water bodies, the so-called emerging pollutants (EP), that can affect both flora and fauna and human health [1]: they include products used daily in households, industry, pharmaceuticals and personal care products, gasoline additives, plasticizers and microplastics [2]. Two main issues of EP are their dynamic character, which is also connected to the improvement of detection techniques, and the difficulty of removal by conventional wastewater treatment technologies. Moreover, emerging pollutants constitute a threat—even at a trace level—because their real impact on human health is unknown.

Although there are no discharge limits for most EP up to now, the European Commission has drawn up and implemented a watch list containing several chemical contaminants that must be monitored with the aim to generate high-quality data on their concentrations in the aquatic environment and to support the risk assessments that underpin the identification of priority substances [3].

During recent years, electro- and photoelectrochemical processes have demonstrated their capacity to efficiently oxidize many of these compounds. Starting from the early 1980s, research on the electrochemical methods for treated wastewater has grown significantly, and thousands of papers now appear in the literature. Although several tests demonstrate the effectiveness of pollutant removal from synthetic and real matrices, this technology is still far from full-scale applications. Its TRL (technology readiness level) is between 4 (technology validated in the lab) and 5 (technology validated in a relevant environment) [4].

More recently, photoelectrochemical processes in which electrochemical and photochemical processes are combined has attracted increasing interest, thanks to the synergy of the two processes: the application of a bias potential improves the photochemical process and the electrochemical process is more efficient since the photo-potential generated on the semiconductor allows for the depolarizing of the cell. This is why, in the last two decades, the number of articles on photochemical wastewater treatment has quickly increased, and the publication of these articles in specific journals indicates that the technology is moving from the fundamentals to real applications [5]. Nevertheless, the TRL of the photoelectrochemical treatment of wastewater is still at the lab scale, and much more efforts are required to push this technology toward applications in the field.

Thus, this special issue contributes to this context, addressing the synthesis, characterization, and application of new materials, as well as the study of catalytic processes and reaction kinetics.

I thank all of the authors for their valuable contribution to this Special Issue and the editorial team at *Catalyst* for their kindness and constant support.

Funding: This research received no external funding.

Conflicts of Interest: The authors declare no conflict of interest.

References

1. Vasilachi, I.C.; Asiminicesei, D.M.; Fertu, D.I.; Gavrilescu, M. Occurrence and Fate of Emerging Pollutants in Water Environment and Options for Their Removal. *Water* **2021**, *13*, 181. [CrossRef]
2. Murgolo, S.; De Ceglie, C.; Di Iaconi, C.; Mascolo, G. Novel TiO2-based catalysts employed in photocatalysis and photoelectrocatalysis for effective degradation of pharmaceuticals (PhACs) in water: A short review. *Curr. Opin. Green Sustain. Chem.* **2021**, *30*, 100473. [CrossRef]
3. EC (2018) Commission Implementing Decision (EU) 2018/840 of 5 June 2018 establishing a watch list of substances for Union-wide monitoring in the field of water policy pursuant to Directive 2008/105/EC of the European Parliament and of the Council and repealing Commission Implementing Decision (EU) 2015/495. *Off. J. Eur. Union* **2018**, *L 141*, 9–12. Available online: https://eur-lex.europa.eu/eli/dec_impl/2018/840/oj (accessed on 28 April 2021).
4. Lacasa, E.; Cotillas, S.; Saez, C.; Lobato, J.; Cañizares, P.; Rodrigo, M.A. Environmental applications of electrochemical technology. What is needed to enable full-scale applications? *Curr. Opin. Electrochem.* **2019**, *16*, 149–156. [CrossRef]
5. Palmas, S.; Mais, L.; Mascia, M.; Vacca, A. Trend in using TiO2 nanotubes as photoelectrodes in PEC processes for wastewater treatment. *Curr. Opin. Electrochem.* **2021**, *28*, 100699. [CrossRef]

Article

Enhanced Photocatalytic Activity of Au/TiO$_2$ Nanoparticles against Ciprofloxacin

Pedro Martins [1,2,*], Sandro Kappert [3], Hoai Nga Le [3,4], Victor Sebastian [5,6], Klaus Kühn [3], Madalena Alves [1], Luciana Pereira [1], Gianaurelio Cuniberti [3,7,8], Manuel Melle-Franco [9] and Senentxu Lanceros-Méndez [1,10,11,*]

1. Department of Physics/Centre of Biological Engineering, University of Minho, 4710-057 Braga, Portugal; madalena.alves@deb.uminho.pt (M.A.); lucianapereira@deb.uminho.pt (L.P.)
2. IB-S—Institute for Research and Innovation on Bio-Sustainability, University of Minho, 4710-057 Braga, Portugal
3. Institute for Materials Science and Max Bergmann Center of Biomaterials, Technische Universität Dresden Dresden, 01062 Dresden, Germany; sandro_kappert@web.de (S.K.); hnle@nano.tu-dresden.de (H.N.L.); klaus.kuehn@nano.tu-dresden.de (K.K.); gianaurelio.cuniberti@tu-dresden.de (G.C.)
4. Department of Chemical Engineering, Hanoi University of Science and Technology, Hanoi 10000, Vietnam
5. Department of Chemical Engineering, Aragon Institute of Nanoscience (INA), University of Zaragoza, Campus Río Ebro-Edificio I+D, C/Poeta Mariano Esquillor S/N, 50018 Zaragoza, Spain; victorse@unizar.es
6. Networking Research Centre on Bioengineering, Biomaterials and Nanomedicine, Centro de Investigacion Biomédica en Red—Bioengenharía, Biomateriales e Nanomedicina, 28029 Madrid, Spain
7. Dresden Center for Computational Materials Science, Technische Universität Dresden Dresden, 01062 Dresden, Germany
8. Center for Advancing Electronics Dresden, Technische Universität Dresden Dresden, 01062 Dresden, Germany
9. Centro de Investigação em Materiais Cerâmicos e Compósitos, Aveiro Institute of Materials, Department of Chemistry, University of Aveiro, 3810-193 Aveiro, Portugal; manuelmelle.research@gmail.com
10. BCMaterials, Basque Center for Materials, Applications, and Nanostructures, Universidad del País Basco—Euskal Herriko Unibertsitatea, Science Park, 48940 Leioa, Spain
11. IKERBASQUE, Basque Foundation for Science, 48013 Bilbao, Spain
* Correspondence: pamartins@fisica.uminho.pt (P.M.); senentxu.lanceros@bcmaterials.net (S.L.-M.)

Received: 14 January 2020; Accepted: 11 February 2020; Published: 15 February 2020

Abstract: In the last decades, photocatalysis has arisen as a solution to degrade emerging pollutants such as antibiotics. However, the reduced photoactivation of TiO$_2$ under visible radiation constitutes a major drawback because 95% of sunlight radiation is not being used in this process. Thus, it is critical to modify TiO$_2$ nanoparticles to improve the ability to absorb visible radiation from sunlight. This work reports on the synthesis of TiO$_2$ nanoparticles decorated with gold (Au) nanoparticles by deposition-precipitation method for enhanced photocatalytic activity. The produced nanocomposites absorb 40% to 55% more radiation in the visible range than pristine TiO$_2$, the best results being obtained for the synthesis performed at 25 °C and with Au loading of 0.05 to 0.1 wt. %. Experimental tests yielded a higher photocatalytic degradation of 91% and 49% of ciprofloxacin (5 mg/L) under UV and visible radiation, correspondingly. Computational modeling supports the experimental results, showing the ability of Au to bind TiO$_2$ anatase surfaces, the relevant role of Au transferring electrons, and the high affinity of ciprofloxacin to both Au and TiO$_2$ surfaces. Hence, the present work represents a reliable approach to produce efficient photocatalytic materials and an overall contribution in the development of high-performance Au/TiO$_2$ photocatalytic nanostructures through the optimization of the synthesis parameters, photocatalytic conditions, and computational modeling.

Keywords: Au-TiO$_2$; antibiotics; emergent contaminants; nanocatalyst; photocatalysis; GFN-xTB

1. Introduction

The resilience of specific emerging pollutants such as pharmaceuticals to the traditional wastewater treatments makes them spread in variable concentrations in surface and groundwater [1]. Dissemination of antibiotics in nature is one of the most significant environmental concerns as they affect biological metabolism and induce the presence of bacterial resistance among drinking water sources [2]. Photocatalysis has received considerable attention from the scientific community as a possible solution to degrade these compounds [3,4].

Typically, the photocatalytic process takes place when a catalyst is UV irradiated and electron-hole pairs are created that will react with H_2O, OH^-, and O_2 to generate oxidizing species such as the hydroxyl radical (OH•), superoxide radical anions (O_2•−), and hydrogen peroxide (H_2O_2). These species will initiate a series of reactions that will degrade pollutants into harmless compounds (e.g., CO_2 and H_2O).

Photocatalysis presents several advantages when compared with other methods, such as the low cost, and the eco-friendly and straightforward processing conditions [5,6]. Many photocatalysts have been reported in the last decades [7,8]. Among them, titanium dioxide (TiO_2) is the most studied and applied in photocatalysis, mainly because of its remarkable optical and oxidizing properties, superhydrophilicity, chemical stability, and durability [9,10]. Despite the compelling advantages of TiO_2, there are also some drawbacks. One of the main hurdles is the low spectral activation of TiO_2, caused by its wide bandgap (3.0–3.2 eV) excitation that only occurs under radiation in the UV or near the UV region (410–387 nm) [11].

For this reason, solar radiation cannot be efficiently used because only less than 5% of this radiation corresponds to UV [3]. Additionally, the process becomes less cost effective as the UV lamps are required to provide the radiation. Another limitation is the electron-hole pair recombination that decreases the photocatalytic efficiency [12,13].

The research developed in the last decades has been mainly devoted to surpassing those limitations by producing new and more efficient photocatalytic materials. Strategies for metallic and nonmetallic doping, co-doping [14,15], dye sensitization, semiconductor combination, co-catalyst loading, and nanocomposite materials [16,17] have been used and tested. These approaches allow us to reduce the electron-hole recombination rate and enhance the absorption of visible radiation of TiO_2 by introducing intermediate energy levels inside the bandgap [18]. In this scope, several works have reported the functionalization of TiO_2 nanoparticles surfaces with metals such as Au [19], Cu [20], Co [21], and Ag [22]. When irradiated, noble metals nanoparticles at the TiO_2 surface can receive electrons and prevent the recombination of the photo-generated electron-hole pairs [23,24].

Metals such as Au and Ag can increase visible light absorption due to the surface plasmon resonance effect [25,26]. Gold (Au) nanoparticles have attracted considerable attention, mainly because they possess exceptional stability, nontoxicity, and biocompatibility [3]. Their properties are highly dependent on the size and shape of the nanoparticles, allowing a broad range of applications [27,28]. For instance, the literature shows that gold nanoparticles in the range of 5 to 10 nm present an enhanced catalytic activity [29,30]. In this sense, some works focused on the photocatalytic activity of Au/TiO_2 nanocomposite have been published, including interesting review articles [3,29,31].

Different physical-chemical techniques have been exploited to produce Au/TiO_2 nanocomposites with enhanced catalytic properties. For instance, chemical vapor deposition [32], sol-gel [33], spray pyrolysis [34], electrophoretic approach [35], deposition-precipitation (DP) [36], deposition-precipitation using urea [37], impregnation [38], hybridization [39], and surface functionalization [40], among others [41,42]. However, many of these techniques are time-consuming, and few of them have focused on the optimization of the nanocomposite and the computational modeling of its nanostructure. Thus, this work focused on the optimization of a DP, converting the Au/TiO_2 nanocomposite production into a cost-effective and straightforward technique, with enhanced photocatalytic activity, under UV and visible radiation. The method optimization aims for cost reduction, using the lowest Au loading that endows visible spectra photocatalytic activity to the nanocomposite. The computational studies provide further information about the electronic mechanism behind the enhanced photocatalytic activity of the Au/TiO$_2$ nanocomposite, as well as the interaction with the target compound.

The target compound is the fluoroquinolone ciprofloxacin (CIP) (chemical formula in Supplementary Material, Figure S1), belonging to a class of synthetic broad-spectrum antibiotics [43], which is mostly used in medicine (e.g., tuberculosis, pneumonia, or digestive disorders). It is also one of the most prescribed fluoroquinolones in the world and studies has shown its presence in potable water and wastewater, as well as in sewage sludge at variable concentrations from milligrams to nanograms per liter [2,44].

In this work, photocatalytic efficiency during the degradation of CIP under UV and visible illumination was assessed. To the best of our knowledge, this is the first work that combines an optimization process of Au/TiO$_2$ nanocomposite with photocatalytic experiments for CIP degradation and computational modeling that addresses the interaction between Au and TiO$_2$ nanoparticles, as well as the interaction of CIP with the produced nanocomposites.

2. Results and Discussion

2.1. Nanocomposite Characterization

The Au/TiO$_2$ nanocomposites were produced by nanoprecipitation method, and the temperature (25, 60, and 80 °C) and the Au loading (ranging from 0.025 to 0.5 wt. %) were changed to understand how these parameters affect the morphology of the nanocomposites and relate it to the photocatalytic efficiency. In this sense, scanning transmission electron microscopy-high-angle annular dark-field imaging (STEM-HAADF) analysis was performed, and the micrographs of the different nanocomposites are displayed in Figure 1.

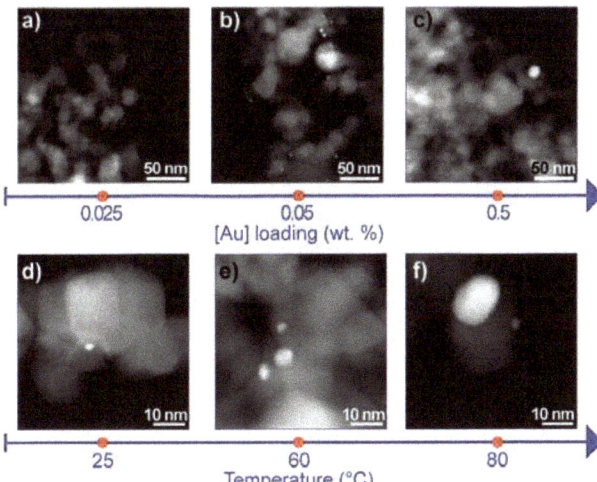

Figure 1. Scanning transmission electron microscopy-high-angle annular dark-field imaging micrographs of Au/TiO$_2$ nanocomposites synthesized with different Au loadings at 60 °C (a–c), and Au/TiO$_2$ nanocomposites obtained at different temperatures with an Au loading of 0.05 wt. % (d–f).

The Au loading study (Figure 1a–c) was assessed producing different nanocomposites using the same experimental conditions (temperature = 60 °C) and changing the loading of gold exclusively, from 0.025 to 0.5 wt. %. The STEM-HAADF micrographs show that for the sample with 0.025 wt. % of Au (Figure 1a), the presence of Au nanoparticles over the surface of the TiO$_2$ nanoparticles was almost inexistent (Figure 1a). With the increase of Au loading to 0.05 wt. % (Figure 1b), it was possible to observe a homogeneous distribution of predominantly small Au nanoparticles (bright contrast nanoparticles below 5 nm in diameter) over the TiO$_2$ nanoparticles. Similar results were obtained for 0.1 wt. % (data not shown). For the concentrations of 0.25 and 0.5 wt. % (Figure 1a–c), agglomerates of Au over theTiO$_2$ nanoparticles (brightest areas of the micrograph) were identified as well as large Au nanoparticles. Analogously, the effect of temperature on the synthesis product was also performed maintaining all the synthesis parameters (Au loading = 0.05 wt. % yielded a homogeneous distribution and size of Au nanoparticles) and changing the temperature of the different samples. STEM-HAADF images (Figure 1d–f) indicate that although the used Au loading was the same in the three temperatures tested when the nanocomposite was synthesized at 80 °C, larger Au nanoparticles appeared more frequently on the nanocomposite (Figure 1f). Conversely, at lower temperatures (25 and 60 °C), the Au nanoparticles size was smaller (Figure 1d,e).

The study of the effect of Au loading and temperature in the nanocomposites morphology indicates that the samples produced at 60 °C and with an Au loading of 0.05 wt. % possessed the more homogeneous distribution and size of Au nanoparticles. In this way, a more detailed STEM-HAADF analysis (Figure 2) was performed on this sample. Figure 2a,b reveal a homogeneous dispersion of Au nanoparticles (white arrows) over TiO$_2$ nanoparticles' surface. The representation of the sphere-like shape of Au nanoparticles in Figure 2c, where an high-resolution scanning transmission electron microscopy – high-angle annular dark field shows that single-crystal nanoparticles with high crystallinity were produced by the proposed method. Size distribution, ranging from 1 to 7 nm, and the average size of 3.2 ± 1.13 nm (Figure 2d), were quantified using Image J software applied to 400 nanoparticles. The size distribution of Au nanoparticles for synthesis at 25 °C and 80 °C is provided in Supplementary Material (Figure S2). All the images show Au nanoparticles with similar sizes, which is in good agreement with the size distribution histogram that presents a sharp size distribution.

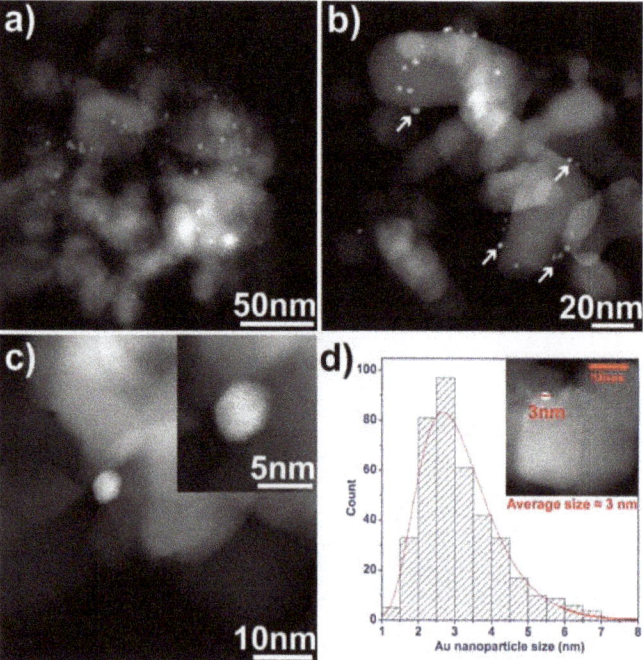

Figure 2. STEM-HAADF micrographs of Au/TiO$_2$ nanocomposites (produced at 60 °C and Au loading of 0.05 wt. %) at different scales (**a**) 50 and (**b**) 200 nm; detail of Au nanoparticle over TiO$_2$ nanoparticles' surface and single Au nanoparticle amplification (inset) (**c**); size distribution of 400 Au nanoparticles with the respective average size (**d**).

The STEM-HAADF- energy-dispersive X-ray spectroscopy (EDX) measurements allowed us to identify the elements present in the Au/TiO$_2$ sample in two different points, 1 and 2 (signaled in Figure 3a). STEM-HAADF-EDX spectra in Figure 3b in point 1 indicate the presence of Au and Cu (copper), which can be respectively addressed to Au nanoparticles and copper grid. In point 2, the signatures of Ti (titanium) and O (oxygen) were identified, corresponding to TiO$_2$ nanoparticles. Thus, EDX measurements confirmed the presence of all the elements of the Au/TiO$_2$ nanocomposite.

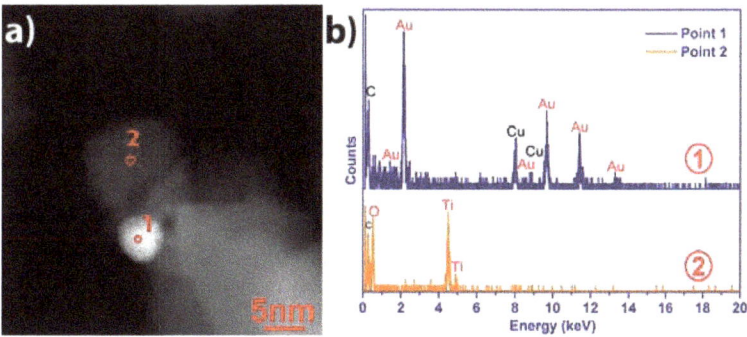

Figure 3. The STEM-HAADF- energy-dispersive X-ray spectroscopy (EDX) image of Au/TiO$_2$ nanocomposites with the identification of the measured points: Au (1) and TiO$_2$ (2) (**a**), EDX spectra with elemental identification (Au, Ti, O, and C) for points 1 and 2 (**b**). The Au/TiO$_2$ nanocomposite synthesized at 60 °C and with an Au loading of 0.05 wt. % was used.

X-ray diffraction was performed to assess the crystal structure of the pure TiO$_2$ nanoparticles and Au/TiO$_2$ nanocomposite, Figure 4a. Both samples show the typical reflexes from anatase (25.3°, 37.8°, and 48.0°) and rutile (27.49°). There was no significant difference between the intensities or positions of the reflexes from both samples. Moreover, no reflexes of Au were detected, which can be explained by the low amount of Au present in the nanocomposite (below detection limit). Figure 4b shows the study of hydrodynamic size for TiO$_2$ and Au/TiO$_2$ nanocomposites obtained by dynamic light scattering (DLS). The results indicated nanoparticles diameters of 1023 nm and 342 nm, for the pristine TiO$_2$ and the Au/TiO$_2$ nanocomposites, respectively. The results suggest that the presence of Au nanoparticles over TiO$_2$ nanoparticles surface may prevent the formation of nanoparticles' aggregates. On the other hand, the size distribution was broader for the nanocomposites regarding the pristine TiO$_2$. Previous work equally showed that the presence of erbium (Er) on TiO$_2$ nanoparticles contributed to reducing the hydrodynamic size when compared with bare TiO$_2$ [15].

The zeta potential was studied at different pH values (3, 5, 7, 9, and 11) for TiO$_2$ and Au/TiO$_2$ samples and the results are displayed in Figure 4c. The pristine and the Au/TiO$_2$ presented very similar profiles, with higher zeta potential values ≈ |20| mV for pH below 3 and 9. These data were in good agreement with the literature [45], with positive zeta potential values for acidic conditions and negative values for basic pH. The more significant difference between the two samples occurred at pH = 7, with the nanocomposite presenting higher zeta potential values than the pure TiO$_2$. Higher zeta potential values mean that nanoparticles possess higher periphery surface charge, which promotes nanoparticles' repulsions, avoiding aggregates' formation and enhanced stability [46]. In this context, and relating it with DLS-obtained results, the smaller hydrodynamic size was probably obtained for the Au/TiO$_2$ because repulsions endowed by Au on TiO$_2$ nanoparticles surface prevented the formation of the aggregates.

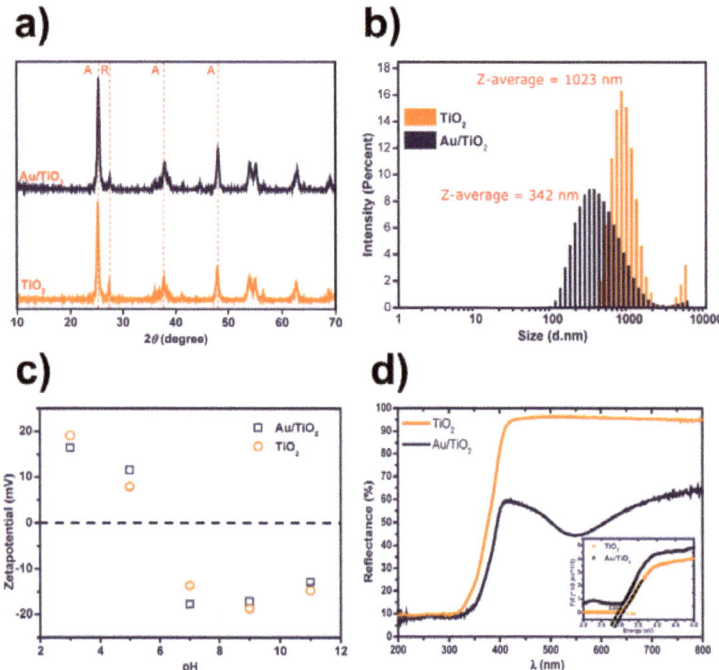

Figure 4. X-ray diffraction reflexes of pristine TiO_2 and Au/TiO_2 nanocomposite and identification of the representative peaks for anatase (A) and rutile (R) phases (**a**); dynamic light scattering, intensity size distribution of the pristine TiO_2 and the Au/TiO_2 nanocomposite and respective Z-average hydrodynamic size (**b**); zeta potential measurements, performed at different pHs (3, 5, 7, 9, and 11) for pristine TiO_2 nanoparticles and Au/TiO_2 nanocomposite (**c**); UV–vis reflectance spectra of pristine TiO_2 and Au/TiO_2 and (inset) the estimation of the bandgap for both samples at $(F(R))1/2 = 0$ (**d**). The Au/TiO_2 nanocomposite synthesized at 60 °C and with an Au loading of 0.05 wt. % was used.

To understand the differences in the photocatalytic performance of TiO_2 and Au/TiO_2 nanocomposite, the optical properties of these materials were studied by UV-visible diffuse reflectance spectra (DRS), depicted in Figure 4d. In the visible range (400–800 nm), the pure TiO_2 nanoparticles reflect the radiation almost entirely ($\approx$ 95%). However, the nanocomposite displays reflectance below 64% for the same range. Additionally, a minimum reflectance ($\approx$ 44%) was obtained at 545 nm, indicating a maximum of absorbance band that can be associated with the surface plasmon of Au nanoparticles, typically in the wavelength range between 520 and 560 nm [47,48]. These results show that the nanocomposite presented a broad absorbance spectrum when compared to the pristine TiO_2 nanoparticles, which is also consistent with the purple/pink color exhibited by the produced nanocomposite. In the ultraviolet range (200–400 nm), both samples showed similar behavior.

From DRS spectra it was possible to estimate the band gap, shown in the inset graph of Figure 4d, for pure TiO_2 and Au/TiO_2 nanocomposite was converting the reflectance to Kubelka–Munk units through Equation (1) and Equation (2). The obtained values show that the nanocomposites possessed a lower bandgap (2.84 eV) than the pristine TiO_2 nanoparticles (2.96 eV). The decrease of the bandgap in Au/TiO_2 was related to the shift absorption to longer wavelengths. Similar results have been reported in the literature [49,50].

2.2. Nanocomposites' Optimization and Photocatalytic Experiments

The photocatalytic activity of all the produced Au/TiO$_2$ nanocomposites was assessed by monitoring the degradation of CIP under artificial UV and visible irradiations. Process conditions were varied depending on the studying purposes.

Nanocomposite Optimization

As gold is a noble metal, cost-effectiveness should be considered, and the amount of gold used in the nanocomposite is one of the most paramount parameters. In this study, Au loading was varied by using different concentrations of the gold precursor. The tested Au loadings were 0.025, 0.05, 0.1, 0.25, and 0.5 wt. %. These nanocomposites were employed for the photocatalytic degradation of CIP under both UV and simulated visible radiation.

Figure 5a shows the data of photocatalytic experiments under UV light. Accordingly, all produced samples and the pristine TiO$_2$ used as a control showed photocatalytic activity, proven by the decrease of CIP concentration along with the irradiation time. As confirmed by the diffuse reflectance spectroscopy (Section 2.1), the bandgap of the nanocomposites was 2.84 eV, corresponding to the wavelength of 437 nm. Here, the used UV lamp had the mode wavelength of 365 nm, which was shorter than the bandgap. It means that the photon energy was adequate to excite the photocatalytic materials, and photocatalytic reaction occurred in all experiments. Pristine TiO$_2$ was compared with the synthesized photocatalysts. After 30 min, 77% of CIP was degraded in the presence of pristine TiO$_2$, whereas higher degradation of 80–90% was achieved in the same time of irradiation, using the synthesized photocatalysts, 0.05 wt. %. This efficiency can be assigned to the presence of gold particles on the surface of the photocatalysts, confirmed by the TEM and EDX characterization (Section 2.1). The further quantitative inspection was obtained using the Langmuir–Hinshelwood kinetics (Equation (3)), and data are shown in Table 1. The apparent reaction rate constant k of the experiment with the bare TiO$_2$ was found to be 0.047 min^{-1}, while the decoration with gold particles improved the photocatalytic activity by 2–3 times. As predicted, in the presence of gold, the excited electrons may be conducted to the gold particles, and the electron-hole recombination may be reduced, which prolongs the lifetime of generated holes [51,52].

Consequently, the photocatalytic activity of the composites increased. Additionally, the increase of the used chloroauric acid concentration might induce a more significant number of gold particles distributed on the TiO$_2$ surface. In other words, the number of electron absorption centers was increased, which explains the increase of k from 0.078 to 0.131 min^{-1} when increasing the Au loading from 0.025 to 0.5 wt. %. However, the further increase in the Au loading caused a decrease in k. These results can be addressed to the loss of photocatalytic active sites on the surface of TiO$_2$ nanoparticles. Based on the TEM images shown in Figure 1, when the Au loading was very high both the amount and the size of Au nanoparticles over the surface of TiO$_2$ nanoparticles were larger, which contributed to a reduction of the adsorption and probably to mitigate the radiation absorbance by the catalytic nanoparticles. Together, these limitations contributed to reducing the photocatalytic efficiency of the nanocomposite towards the samples with lower amounts of Au and demonstrated the relevance of optimizing the Au loading.

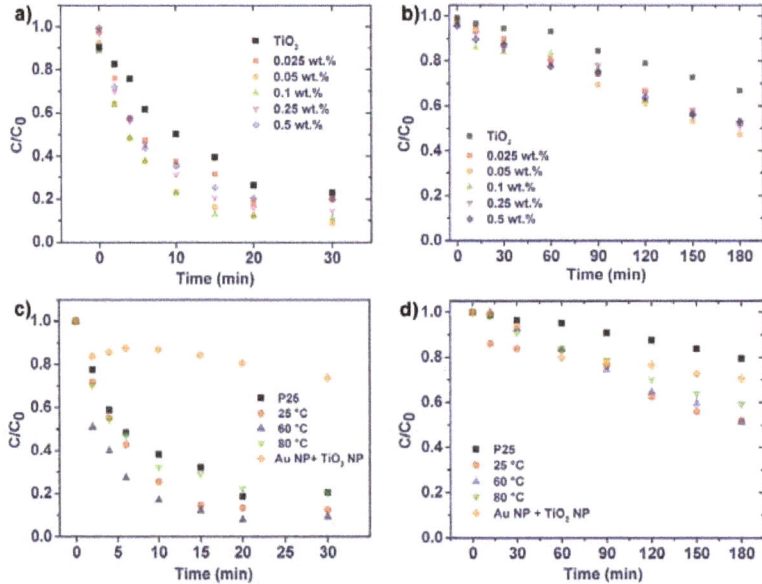

Figure 5. Photocatalytic degradation of ciprofloxacin (5 mg/L) with bare TiO_2 and Au/TiO_2 nanocomposite with different Au concentrations under 30 and 180 minutes of UV (**a**) and visible (**b**) radiation. The degradation with bare TiO_2 and Au/TiO_2 nanocomposites synthesized at different temperatures and Au loading of 0.05 wt. % under 30 and 180 minutes of UV (**c**) and simulated visible light radiation (**d**), respectively.

Table 1. Apparent reaction rates (k) for photocatalytic degradation of ciprofloxacin (CIP) (5 mg/L) with bare TiO_2 and Au/TiO_2 nanocomposite with different Au loadings, over 30 and 180 minutes of UV and simulated visible radiation, respectively.

Au loading (wt. %)	UV k (min^{-1})	Simulated Visible k (h^{-1})
0	0.047	0.073
0.025	0.078	0.211
0.05	0.099	0.242
0.1	0.131	0.211
0.25	0.089	0.195
0.5	0.076	0.202

The photocatalytic assays performed under visible illumination are shown in Figure 6b. Regarding these assays, it is essential first to mention the controls (Supplementary Material, Figures S3 and S4), which have shown that the CIP solution was stable under simulated visible radiation, demonstrating its photostability. Moreover, another control was performed by adding the Au/TiO_2 nanocomposites to CIP solution in the dark for 180 minutes. In this case, approximately 11% of CIP was removed from the solution by adsorption to the Au/TiO_2 nanocomposites.

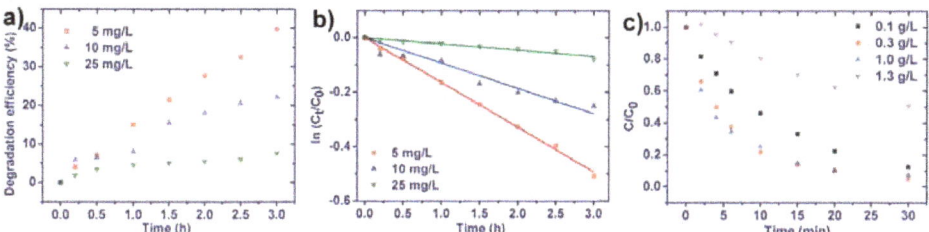

Figure 6. Degradation efficiency (%) (**a**) and ln (C/C0) vs. time (**b**) for different initial ciprofloxacin concentrations (5, 10, and 25 mg/L), using Au/TiO$_2$ nanocomposites produced at 60 °C and with an Au loading of 0.05 wt. %, under 3 hours of simulated visible radiation. Photocatalytic degradation of ciprofloxacin (5 mg/L) in 45 mL of aqueous solution with different Au/TiO$_2$ concentrations (0.1, 0.3, 1.0, and 1.3 g/L). The Au/TiO$_2$ nanocomposite synthesized at 60 °C and with an Au loading of 0.05 wt. % was used. The tests were performed over 30 minutes under UV irradiation (**c**).

With the information from controls, it is possible to understand the photocatalytic efficiency of the tested materials better. Similarly, to the UV light experiments, the degradation rates of all produced nanocomposites were faster than that with the bare TiO$_2$. TiO$_2$ could remove ≈ 33% of CIP after 180 min of simulated visible irradiation. This CIP removal may be assigned to adsorption, confirmed by controls performed in the dark (as above mentioned). Additionally, the sun simulator device had a small percentage (≈ 3%) of UV radiation (to mimic sunlight radiation). This radiation can induce a low photocatalytic activity on bare TiO$_2$, which, together with the adsorption of CIP, is responsible for its removal from the solution.

More importantly, the decoration of gold particles on the TiO$_2$ surface resulted in the faster degradation rate of CIP under visible radiation. The bandgap of the composites was lowered, from 2.96 eV to 2.84 eV (Section 2.1). Similar results were obtained for methylene blue degradation using Au/TiO$_2$ nanoparticles. The authors obtained higher degradation efficiencies and ability to use visible radiation [37]. Thus, the materials could absorb the longer wavelength in the visible range (up to 437 nm). The reaction rates' constant increased from 0.073 h^{-1}, without Au, to 0.195–0.224 h^{-1}, with different Au loadings (Table 1).

The obtained results, for UV and visible radiation, confirmed that the photocatalytic efficiency of the TiO$_2$ nanoparticles was enhanced with the Au loading, until a specific plateau. When the Au loading was higher than 0.1 and 0.05 wt. %, respectively, for UV and visible radiation, the gold nanoparticles can block the surface-active sites of TiO$_2$ nanoparticles [53,54]. Furthermore, an excessive amount of Au nanoparticles can play as recombination centers for photo-induced electrons and holes. Both situations can contribute to a significant reduction of pollutant adsorption and, consequently, the photocatalytic efficiency [55]. The remaining assays of this study will be performed with an Au loading of 0.05 wt. %.

Another critical parameter that is worth to stress and study is temperature, which can affect the surface charge phenomenon and the dispersity of the TiO$_2$ particles in the solution during the synthesis. It can also influence the nucleation and the gold particles' crystal growth on the TiO$_2$ nanoparticle surface. In this study, the synthesis was operated at 25, 60, and 80 °C, and the photocatalytic degradation of CIP, with the nanocomposites produced at different temperatures, was performed under UV and visible radiation (Figure 5c,d and Table 2). Regardless of the synthesis temperature, the photocatalytic activity of the nanocomposites (Au loading = 0.05 wt. %) was equal or higher than that of the bare TiO$_2$. Here, the synthesis at the room and medium temperatures (25 and 60 °C) yielded the more efficient photocatalytic materials, for UV and visible radiation, towards higher temperature synthesis (80 °C).

Table 2. Apparent reaction rates (k) for photocatalytic degradation of CIP (5 mg/L) with bare TiO_2 and Au/TiO_2 nanocomposite synthesized at different temperatures, over 30 and 180 minutes of UV and simulated visible radiation, respectively. The Au loading of 0.05 wt. % was used for the tested materials.

Temperature (°C)	UV		Simulated Visible	
	k (min^{-1})	Degradation (%)	k (h^{-1})	Degradation (%)
TiO_2	0.66	80	0.073	20
25	0.131	88	0.221	48
60	0.117	91	0.226	49
80	0.047	80	0.176	41

For both types of radiation, the sample obtained at 60 °C presented higher degradation efficiencies (Table 2), 91% and 49% of CIP degradation under UV (30 min) and visible radiation (180 min), respectively. On the other hand, the synthesis performed at 80 °C revealed lower degradations rates of 80% and 40% for UV and visible radiation, respectively. Other works have reported that higher temperatures accelerate the reduction process and yield broader Au nanoparticles size distributions [56].

In this context, when the synthesis occurred at 80 °C, the size of Au nanoparticles produced was larger than the sizes obtained with 25 °C and 60 °C (in good agreement with STEM-HAADF micrographs, Figure 1). Similarly to what happened with the Au loading, when the amount of Au on the surface of TiO_2 was too high, the active sites were blocked and the pollutant adsorption can be limited. Compared with bare TiO_2, these results corresponded to a degradation efficiency increase of approximately 13% and 145% for UV and visible radiation, respectively.

Both under UV and visible radiation, another control was performed (Figure 5c,d) by testing single Au nanoparticles at the very same amount of Au (corresponding to 0.05 wt. % obtained at 60 °C) and TiO_2 nanoparticles on CIP degradation. The results confirmed that the photocatalytic efficiency obtained by the nanocomposites should be assigned to the interface between Au and the TiO_2 surface.

2.3. Photocatalytic Degradation

The rate of photocatalytic degradation depends on the availability of the catalyst surface for the photo-generation of electron-hole pairs that produce hydroxyl radicals. Thus, in these experiments, the amount of catalyst was kept constant, and the number of hydroxyl radicals generated remained the same, while CIP concentration increased. The influence of CIP initial concentration of 5, 10, and 25 mg/L was studied under visible irradiation. It was observed that the CIP concentration impacted by the degradation rate and efficiency (Figure 6). With the lowest CIP concentration, 40% of CIP degradation was obtained after 30 min. With the increase of concentrations by 2 and 5 times, the efficiencies achieved were 22% and 8%, respectively. In these tests, while using the photocatalyst concentration of 0.3 g/L, the adsorption of the CIP on the Au/TiO_2 nanoparticles surface might be halted due to surface saturation. Additionally, the presence of organic compounds such as CIP can generate an increased number of intermediates and products, which will compete with CIP for adsorption on the photocatalyst surface [57]. This competition caused a lower reaction rate for high CIP concentration. The following assays, focused on the photocatalytic activity of the produced nanocomposites, were performed using the lowest CIP concentration, 5 mg/L.

In short, the ratio between hydroxyl radical/CIP molecules decreased with higher concentrations, causing lower photocatalytic activity. Moreover, higher CIP concentrations may also reduce radiation harvesting by TiO_2 nanoparticles surface, which will also contribute to decreasing the number of hydroxyl radicals formed. Figure 6b displays the plot of ln (C/C0) vs. time at different initial CIP concentrations. Linear plots were observed, and the R^2 values were higher than 0.9, confirming that the photocatalytic degradation of CIP obeyed pseudo first-order kinetics.

The optimal photocatalyst concentration was assessed through degradation of CIP with the different amounts of Au/TiO_2 nanocomposites, from 5 to 60 mg, which corresponded to a photocatalyst concentration of 0.1 and 1.3 g/L, respectively. Experimental results are shown in Figure 6c.

In general, with the photocatalytic concentrations of 0.1–1.0 g/L, ≈ 90% of the CIP in solution was degraded, while with the higher concentration of 1.3 g/L, only 50% CIP was degraded after exposure to the same UV irradiation time. At the lowest photocatalytic concentrations of 0.1–0.3 g/L, the photocatalytic degradation increased significantly with the amount of nanocomposite. Indeed, the increased amount of photocatalyst resulted in higher surface coverage, owing to the highest number of active sites [58]. However, when increasing the concentration to 1.0 g/L, the degradation rate remained unchanged. With the highest concentration of Au/TiO2, 1.3 g/L, the degradation was slowest likely because of excessive turbidity that induced light extinguishment after penetrating a short distance from the illuminated surface [59]. Most of the light might be extinguished after penetrating a short distance from the illuminated surface of the suspension. Thus, photocatalytic particles in an inner region could not be activated. The result agrees with other reports [58,60]. For this reason, the concentration of 0.3 g/L (15 mg in 45 mL), which yielded the highest photocatalytic efficiency, was chosen for the following photocatalytic activity assays.

The reproducibility of the nanocomposites efficiency was also tested using three independent syntheses, performed under the same conditions. The produced samples were then used in the photocatalytic degradation of CIP under visible radiation (results in the Supplementary Material, Figure S5). The apparent reaction rate constant of the three experiments fluctuated around the value of 0.219 ± 0.022 min^{-1}. The standard deviation of 10% proved the reproducibility of the method to enhance the photocatalytic activity of pure TiO_2 nanoparticles and endow them with visible light activity. It is also important to clarify that the Au concentrations used in this study were nominal, as no inductively coupled plasma atomic emission spectroscopy (ICP-AES), or similar characterization, was performed. However, given the considerable reproducibility of the method, the putative loss of gold would by similar for all the Au concentrations tested, making them comparable.

3. Computational Modeling: Gold on Titanium Dioxide and Charge Transfer

A computational study was performed to rationalize the effect of gold nanoparticles on TiO_2. The GFN-xTB (Geometry, Frequency, Noncovalent, eXtended Tight-Binding) was used. GFN-xTB is a new semiempirical method developed by Grimme et al. [61] that allows computing efficiently systems with thousands of atoms. The GFN-xTB software used (version 5.4.6) did not allow us to compute systems with periodic boundary conditions, so a finite system composed of a gold nanoparticle adsorbed on a larger TiO_2 nanoparticle was used. We chose a cuboctahedral $(TiO_2)_{97}$ anatase nanoparticle, which was found to produce bulk-like electronic properties [62] and had two large, equal, flat surfaces of ~ 1.2×1.5 nm^2 on which a cuboctahedral $(Au_{55})^{-3}$, 10 nm diameter, gold nanoparticle was adsorbed [63]. Several adsorption modes were possible, yet an exhaustive search of these was beyond the scope of this study. We positioned the gold nanoparticle on four, arbitrary, different orientations so that, in all cases, one of the faces of $(Au_{55})^{-3}$ was parallel to the anatase flat surface and minimized. In this minimization, the TiO2 coordinates were held fixed while the Au first neighbors' distances were constrained harmonically to an equilibrium value to force the gold nanoparticle to keep its initial shape while retaining some flexibility. Two different pH conditions were considered: (1) A neutral/basic pH represented by the bare $(TiO_2)_{97}$ anatase nanoparticles (i.e., without protonation), and (2) an acidic pH, where the eight under-coordinated oxygen atoms in the anatase nanoparticle were protonated, $(TiO_2)_{97}H_8^{+8}$, which should represent better the experimental conditions (Figure 7).

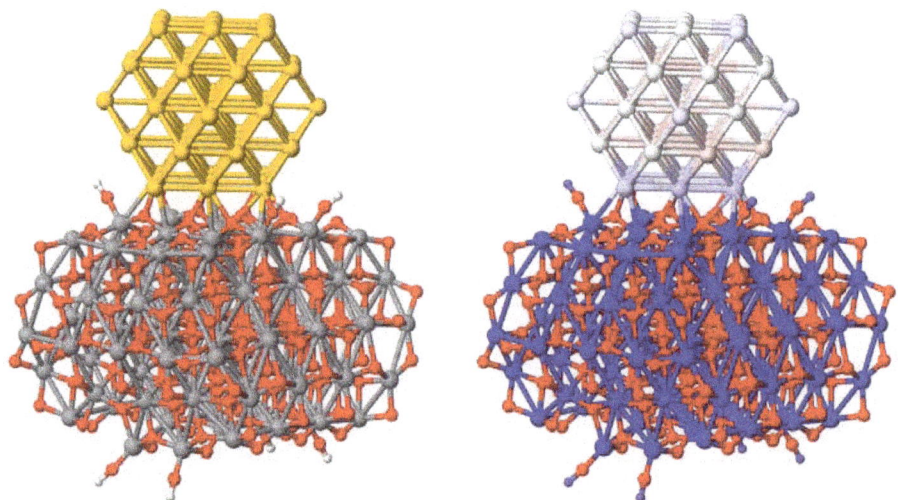

Figure 7. The $(Au_{55}(TiO_2)_{97}H_8)^{+5}$ with standard atom coloring (left) with color-rendered atomic charges (right), so that red represents negatively charged atoms, blue positively charged atoms, and white neutral atoms.

To study the charge transfer between $(Au_{55})^{-3}$ and $(TiO_2)_{97}$ and $((TiO_2)_{97}H_8)^{+8}$, we computed atomic charges specifically suited for condensed phases [64]. The analysis of the charges in $(Au_{55}(TiO_2)_{97})^{-3}$ and $(Au_{55}(TiO_2)_{97}H_8)^{+5}$ (Figure 7) show that the Au nanoparticle transferred electrons to the anatase, namely 4.7 electrons for neutral TiO_2 and 4.3 electrons for the protonated TiO_2. Most of this charge was transferred directly to Ti atoms, namely 3 and 3.3 electrons for the neutral and protonated case, respectively. So, the Au acted as an electron donor populating the Ti(d) states that were responsible for the photocatalytic activity of the material. Consequently, Au may increase the catalytic activity of the composite material through interfacial electron transfer. Interestingly, the anatase surface also polarized the gold nanoparticle so that all atoms in direct contact with TiO_2 were more oxidized, i.e., had larger positive charges, see Figures 7 and 8.

This effect was observed in all cases, for neutral and acidic pH and when the harmonic constraint on Au atoms was released, which indicates that the observed charge transfer is a fundamental process of the Au-TiO_2 interface. This corroborated the experimental controls shown in Figure 5c,d, showing that separation of Au and TiO_2 nanoparticles yielded lower efficiencies. Moreover, this finding also fit previous DFT (density functional theory) calculations which found that an Au nanorod on a rutile TiO_2 (110) surface might act as an activator for molecular oxygen through charge transfer to nearby Ti^{+4} atoms [65]. Mechanistically, the presence of Au activated superficial Ti^{+4} atoms nearby for catalysis via direct charge transfer, which rationalized our experimental observation that lower Au loading and small gold nanoparticles had larger catalytic activity than larger loadings and nanoparticles on TiO_2.

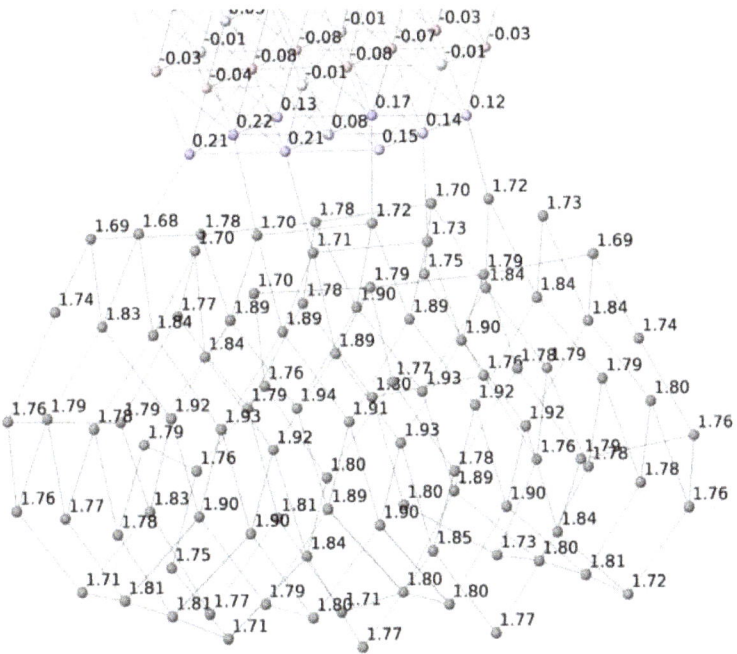

Figure 8. The $(Au_{55}(TiO_2)_{97}H_8)^{+5}$ with only the titanium subnetwork and gold atoms with their corresponding point charges.

Adsorption of Ciprofloxacin on $Au_{55}(TiO_2)_{97}H_8^{+58}$

We also explored the different energetics of a CIP molecule interacting with $(Au_{55}(TiO_2)_{97}H_8)^{+58}$ on four different adsorption sites (Figure 9). All the adsorption modes were found to be binding, i.e., exothermic, with energies ranging between 0.7 and 2.4 eV. The strongest binding was observed for the structure with a large contact between the oxygen atom of the carboxylic group and the gold nanoparticle, 2.4 eV. This binding was similar in energy to the adsorption on the anatase clean surface (2.1 eV), which indicates that both processes might be competitive and that CIP molecules might also adsorb near the gold/anatase interface where the Ti atoms were activated through electron transfer.

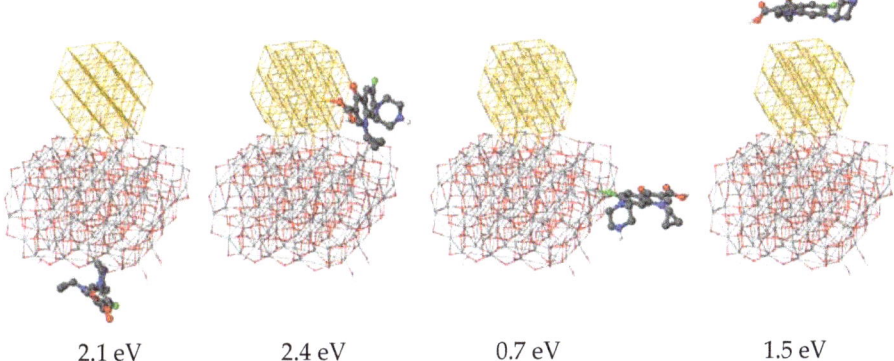

Figure 9. Four possible adsorption geometries of CIP on $(Au_{55}(TiO_2)_{97}H_8)^{+58}$ with corresponding binding energies. The positive binding energies indicate that the process is exothermic.

4. Materials and Methods

4.1. Materials

P25 TiO$_2$ nanoparticles were kindly provided by Evonik (Essen, Alemanha). Gold(III) chloride trihydrate, 99.9% CAS: 16961-25-4 (liquid solution) was purchased from Sigma-Aldrich (St. Louis, Missouri, EUA). Sodium hydroxide (NaOH) was obtained from VWR (Radnor, Pensilvânia, EUA) Millipore Milli-Q-system ultra-pure (UP) water was used in all the experiments.

4.2. Nanocomposite Synthesis

The Au/TiO$_2$ nanocomposites were synthesized, as illustrated in Figure 10, dispersing 200 mg of TiO$_2$-P25 nanoparticles in 40 mL of ultra-pure (UP) water in a sonication bath for 30 min. Afterwards, this solution was placed under agitation in a water bath at different temperatures (25, 60, and 80 °C), using a thermostat to precisely control and stabilize the temperature, avoiding thermal gradients. When the dispersion solution reached the desired temperature, different volumes from the chloroauric solution (10 µL of Gold(III) chloride trihydrate in 100 mL of UP water) were added to achieve the Au loadings of 0.025, 0.05. 0.1, 0.25, and 0.5 wt. %. The solution was then stirred for 10 minutes to achieve a homogeneous distribution of gold precursor solution. Later, several volumes of a 0.1 M sodium hydroxide solution (NaOH) were added dropwise and mixed for 10 minutes to obtain a pH = 9. The solution was then centrifuged at 23,000 rpm, the supernatant discarded, and the nanocomposite pellet redispersed in UP water with the ultrasonication for 1 minute, and this washing procedure was repeated one more time. The last step was to dry the nanocomposite at 80 °C in an oven overnight and grind it with a pestle and mortar to obtain a fine powder.

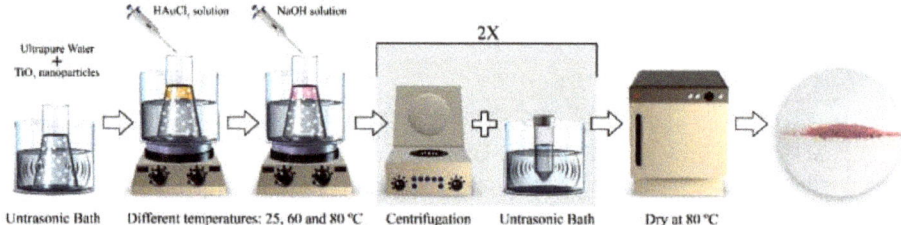

Figure 10. Schematic representation of the main steps to synthesize Au/TiO$_2$ nanocomposites trough nanoprecipitation.

4.3. Characterization

The morphology of the nanocomposites was assessed by transmission electron microscopy (TEM), a Tecnai T20 from FEI (Hillsboro, Oregon, EUA). For the analysis, the nanocomposite samples were sonicated for 5 minutes to achieve good dispersion and afterwards a drop of the suspension was placed on a copper grid and dried at room temperature for the analysis. Particle size histograms were obtained after measuring at least 200 nanoparticles using Image J 1.50i software. Aberration-corrected scanning transmission electron microscopy (Cs-corrected STEM) images were acquired using a high-angle annular dark field detector in an FEI XFEG TITAN (Hillsboro, Oregon, EUA) electron microscope operated at 300 kV equipped with a Spherical Aberration Corrector for Transmission Electron Microscopes (CETCOR) Cs-probe corrector from CEOS Company (Heidelberg, Germany), allowing the formation of an electron probe of 0.08 nm. Elemental analysis was carried out with an EDX (energy-dispersive X-ray spectroscopy) detector, which allows performing EDX experiments in the scanning mode.

The crystallographic phases of the pure TiO$_2$ and the Au/TiO$_2$ nanocomposite were evaluated by X-ray diffraction using a D8 Discover diffractometer with incident Cu Kα (40 kV and 30 mA), from Bruker (Billerica, Massachusetts, EUA).

The average hydrodynamic diameter was assessed by dynamic light scattering (DLS) in a Zetasizer NANO ZS-ZEN3600, Malvern (Malvern Instruments Limited, United Kingdom), equipped with a He–Ne laser (wavelength 633 nm) and backscatter detection (173°). The samples were dispersed (0.1 mg/L) in ultrasonication bath at 22 °C for 30 minutes to avoid aggregates, and each sample was measured 10 times. The zeta (ζ) potential was measured in the same device, and TiO_2 nanoparticles were equally suspended in ultra-pure water and solutions at different pHs (2, 4, 7, 9, and 12) were prepared with HCl (1M) and NaOH (1M) solutions. The results were obtained using the Smoluchowski theory approximation, and each sample was measured 10 times at 22 °C. The manufacturer software (Zetasizer 7.12) was used to assess particles diameter (intensity distribution), the polydispersity index (PDI), and z-potential values.

The optical properties of the pristine TiO_2 and the Au/TiO_2 nanocomposite were assessed by UV–vis reflectance, using a Shimadzu UV-2501-PC (Kyoto, Japan) equipped with an integrating sphere. The spectra were acquired in reflectance, and the bandgap was estimated via the Kubelka–Munk Equation (1) [52] and the Tauc plot represented by Equation (2).

$$F(R) = (1 - R_\infty)^2 / (2R_\infty) \tag{1}$$

where R_∞ (R_{Sample}/R_{BaSO4}) corresponds to the reflectance of the sample and $F(R)$ is the absorbance.

$$[F(R)h\upsilon]^{1/n} \text{ versus } h\upsilon \tag{2}$$

where h is the Planck constant (6.626 × 10^{-19} J), n is the frequency, and n is the sample transition (indirect transition, n = 2) [66].

4.4. Photocatalytic Degradation

The photocatalytic activity of all the produced samples and pristine TiO_2 was assessed by performing (CIP) degradation tests, under artificial ultraviolet (UV) or visible illumination. First, a solution of 5 mg/L of CIP was prepared. The CIP solution was adjusted to pH = 3, to ensure the solubility, by using 0.1 mL hydrochloric acid (HCl) 1 M.

Before the degradation assays (UV or visible radiation), the Au/TiO_2 or P25 nanoparticles were stirred in the dark for 30 min to achieve an adsorption-desorption equilibrium.

The UV degradation of CIP was performed in a chamber with six Philips 8 W mercury fluorescent lamps with the mode wavelength of 365 nm. The suspensions of photocatalysts and CIP were kept stirred in a container under the illumination from the top. The distance between the beaker and the lamp was 13.5 cm, and the intensity coming to the system was 15–17 W/m^2. The samples were irradiated for 30 min.

The visible light tests were performed in a visible chamber fabricated by Ingenieurbüro Mencke & Tegtmeyer GmbH©, Hameln, Alemanha. According to the manufacturer, the visible light spectrum was equivalent to that of the natural solar light. The light source had an intensity of 98 W/m^2. The visible light test was performed similarly to the UV test. Here, the container was placed at 21 cm from the light source, and the samples were irradiated continuously for 180 minutes.

The first photocatalytic activity tests were performed to determine the optimal ratio of CIP/catalyst. For this purpose, 5, 15, 45, and 60 mg of Au/TiO_2 nanocomposite were dispersed in a borosilicate beaker of 80 mL with 45 mL of CIP solution (5 mg/L). The effect of Au loading on the photocatalytic efficiency was also assessed, under UV and simulated visible radiation. The impact of the synthesis temperature (25, 60, and 80 °C) on CIP photocatalytic degradation was equally evaluated using both types of illumination. The photocatalytic reproducibility tests were performed using nanocomposites produced in different batches but under the same synthesis conditions.

The bare TiO_2 nanoparticles were used as controls in all the assays. Additionally, to prove the relevance of the Au/TiO_2 nanocomposites' structure and interface, the photocatalytic degradation of

CIP was assessed using the same amounts of Au and TiO_2 nanoparticles, not as a nanocomposite, but separately added to the solution.

The photocatalytic efficiencies were tested by degrading CIP in aqueous solution under UV and visible radiation and monitoring the maximum absorption peak (277 nm) using a Shimadzu UV-2501PC UV/Vis spectrophotometer. The degradation fit the Langmuir-Hinshelwood model, expressed by Equation (3):

$$C/C_0 = exp^{-kt} \qquad (3)$$

where C_0 and C represent the concentration of the pollutant at time 0 min and at time t, respectively, and k is the first-order rate constant of the reaction.

5. Conclusions

An Au/TiO_2 nanocomposite was produced, characterized, and applied in the photocatalytic degradation of ciprofloxacin (CIP). The characterization results changing the synthesis conditions (temperature and Au loading) indicated that the synthesis performed at 60 °C with the Au loading of 0.05 wt. % yielded the most homogeneous distribution of Au nanoparticles (≈3 nm) over TiO_2 nanoparticles surface, after TEM inspection. Additionally, these samples absorbed more radiation in the visible range (≈66% at 545 nm) and presented a lower bandgap (2.84 eV vs. 2.96 eV from bare TiO_2). The photocatalytic results confirmed that all the manufactured nanocomposites possessed higher photocatalytic efficiency in the UV and simulated visible radiation towards the pristine TiO_2. It was also possible to understand the impact of the synthesis parameters envisaging the optimal photocatalytic efficiency conditions. In this way, with the Au/TiO_2 nanocomposite, it was possible to enhance the photocatalytic degradation efficiency in 13% and 145% under UV and simulated visible light radiation, respectively. The gold nanoislands played a paramount role transferring electrons from Au to the anatase from TiO_2 nanoparticles. Additionally, Au endowed the nanocomposite with the ability to absorb the visible radiation.

Computational modeling supported the experimental data, showing the ability of Au to bind TiO_2 anatase surfaces as well as the relevant role of Au transferring electrons. The fundamental importance of the interface between TiO_2 and Au nanoparticles regarding the enhanced photocatalytic activity was also rationalized. Moreover, models indicated a high affinity of CIP to both Au and TiO_2 surfaces, which favors the adsorption process and consequently may also be cause for enhanced photocatalytic efficiency in the presence of Au nanoparticles.

According to the results obtained through systematic experimental data and modeling results, the simple method herein presented constitutes a reliable approach to produce efficient photocatalytic materials.

Supplementary Materials: The following are available online at http://www.mdpi.com/2073-4344/10/2/234/s1: Figure S1: Size distribution of Au nanoparticles for synthesization at 25 and 80 °C; Figure S2: Photostability of CIP solution under UV; Figure S3: Photostability of CIP solution under visible radiation; Figure S4: Synthesis reproducibility on CIP degradation.

Author Contributions: Conceptualization, P.M. and S.L.-M.; data curation, P.M., S.K., and H.N.L.; formal analysis, H.N.L., L.P., M.M.-F., and S.L.-M.; investigation, P.M., S.K., H.N.L., and M.M.-F.; methodology, P.M., S.K., and V.S.; project administration, S.L.-M.; resources, V.S., M.A., G.C., and S.L.-M.; software, M.M.-F.; supervision, K.K., M.A., G.C., and S.L.-M.; validation, V.S.; visualization, P.M. and M.M.-F.; writing—original draft, P.M. and H.N.L.; writing—review & editing, P.M., M.A., and S.L.-M. All authors have read and agreed to the published version of the manuscript.

Funding: The authors acknowledge funding from the Basque Government Industry Department under the ELKARTEK Program and the Spanish Ministry of Economy and Competitiveness (MINECO) through the project MAT2016-76039-C4-3-R (AEI/FEDER, UE) (including the FEDER financial support). This work was also supported by the Graduate Academy of the Technische Universität Dresden. Centro de Investigacion Biomédica en Red – Bioengenharía, Biomateriales e Nanomedicina (CIBER-BBN) is an initiative funded by the 6th National R&D&i Plan 2008–2011, Iniciativa Ingenio 2010, Consolider Program, and CIBER Actions and financed by the Instituto de Salud Carlos III (Spain) with assistance from the European Regional Development Fund. S. Kappert and H.N. Le acknowledge fruitful discussions with Nadia Licciardello.

Acknowledgments: This work was supported by the Portuguese Foundation for Science and Technology (FCT) in the framework of the strategic projects UID/FIS/04650/2013 by Fundo Europeu de Desenvolvimento Regional (FEDER) funds through the COMPETE 2020—Programa Operacional Competitividade e Internacionalização (POCI) with the reference project POCI-01-0145-FEDER-006941, project PTDC/CTM-ENE/5387/2014, as well as UID/BIO/04469 unit through COMPETE 2020 (POCI-01-0145-FEDER-006684) and BioTecNorte operation (NORTE-01-0145-FEDER-000004) funded by the European Regional Development Fund under the scope of Norte2020—Programa Operacional Regional do Norte. P.M. Martins thanks the FCT for the grant SFRH/BD/98616/2013 and Luciana Pereira for the grant SFRH/BPD/110235/2015. M. Melle-Franco would like to acknowledge support from Centro de Investigação em Materiais Cerâmicos e Compósitos (CICECO)—Aveiro Institute of Materials, POCI-01-0145-FEDER007679 (UID/CTM/50011/2013) and the FCT (IF/00894/2015).

Conflicts of Interest: The authors declare no conflict of interest.

References

1. An, T.; Yang, H.; Song, W.; Li, G.; Luo, H.; Cooper, W.J. Mechanistic considerations for the advanced oxidation treatment of fluoroquinolone pharmaceutical compounds using TiO_2 heterogeneous catalysis. *J. Phys. Chem. A* **2010**, *114*, 2569–2575. [CrossRef]
2. Li, W.C. Occurrence, sources, and fate of pharmaceuticals in aquatic environment and soil. *Environ. Pollut.* **2014**, *187*, 193–201. [CrossRef]
3. Ayati, A.; Ahmadpour, A.; Bamoharram, F.F.; Tanhaei, B.; Mänttäri, M.; Sillanpää, M. A review on catalytic applications of Au/TiO_2 nanoparticles in the removal of water pollutant. *Chemosphere* **2014**, *107*, 163–174. [CrossRef] [PubMed]
4. Jiang, J.-Q.; Ashekuzzaman, S.M. Development of novel inorganic adsorbent for water treatment. *Curr. Opin. Chem. Eng.* **2012**, *1*, 191–199. [CrossRef]
5. Kumar, S.; Ahlawat, W.; Bhanjana, G.; Heydarifard, S.; Nazhad, M.M.; Dilbaghi, N. Nanotechnology-based water treatment strategies. *J. Nanosci. Nanotechnol.* **2014**, *14*, 1838–1858. [CrossRef] [PubMed]
6. Pawar, R.C.; Lee, C.S. *Heterogeneous Nanocomposite-Photocatalysis for Water Purification*; William Andrew: London, UK, 2015.
7. Lu, S.-Y.; Wu, D.; Wang, Q.-L.; Yan, J.; Buekens, A.G.; Cen, K.-F. Photocatalytic decomposition on nano-TiO_2: Destruction of chloroaromatic compounds. *Chemosphere* **2011**, *82*, 1215–1224. [CrossRef]
8. Khaki, M.R.D.; Shafeeyan, M.S.; Raman, A.A.A.; Daud, W.M.A.W. Application of doped photocatalysts for organic pollutant degradation—A review. *J. Environ. Manag.* **2017**, *198*, 78–94. [CrossRef]
9. Fabregat-Santiago, F.; Barea, E.M.; Bisquert, J.; Mor, G.K.; Shankar, K.; Grimes, C.A. High carrier density and capacitance in TiO_2 nanotube arrays induced by electrochemical doping. *J. Am. Chem. Soc.* **2008**, *130*, 11312–11316. [CrossRef]
10. Hoffmann, M.R.; Martin, S.T.; Choi, W.; Bahnemann, D.W. Environmental applications of semiconductor photocatalysis. *Chem. Rev.* **1995**, *95*, 69–96. [CrossRef]
11. Vargas Hernández, J.; Coste, S.; García Murillo, A.; Carrillo Romo, F.; Kassiba, A. Effects of metal doping (Cu, Ag, Eu) on the electronic and optical behavior of nanostructured TiO_2. *J. Alloy. Compd.* **2017**, *710*, 355–363. [CrossRef]
12. Shen, L.; Liang, R.; Wu, L. Strategies for engineering metal-organic frameworks as efficient photocatalysts. *Chin. J. Catal.* **2015**, *36*, 2071–2088. [CrossRef]
13. Ahmad, R.; Ahmad, Z.; Khan, A.U.; Mastoi, N.R.; Aslam, M.; Kim, J. Photocatalytic systems as an advanced environmental remediation: Recent developments, limitations and new avenues for applications. *J. Environ. Chem. Eng.* **2016**, *4*, 4143–4164. [CrossRef]
14. Carneiro, J.T.; Yang, C.-C.; Moma, J.A.; Moulijn, J.A.; Mul, G. How gold deposition affects anatase performance in the photo-catalytic oxidation of cyclohexane. *Catal. Lett.* **2009**, *129*, 12–19. [CrossRef]
15. Martins, P.M.; Gomez, V.; Lopes, A.C.; Tavares, C.J.; Botelho, G.; Irusta, S.; Lanceros-Mendez, S. Improving photocatalytic performance and recyclability by development of Er-doped and Er/Pr-codoped TiO_2/Poly(vinylidene difluoride)–trifluoroethylene composite membranes. *J. Phys. Chem. C* **2014**, *118*, 27944–27953. [CrossRef]
16. Barakat, M.A.; Kumar, R. Photocatalytic activity enhancement of titanium dioxide nanoparticles. In *Photocatalytic Activity Enhancement of Titanium Dioxide Nanoparticles*; Springer: Berlin, Germany, 2016; pp. 1–29.

17. Almeida, N.A.; Martins, P.M.; Teixeira, S.; Lopes da Silva, J.A.; Sencadas, V.; Kühn, K.; Cuniberti, G.; Lanceros-Mendez, S.; Marques, P.A.A.P. TiO$_2$/graphene oxide immobilized in P(VDF-TrFE) electrospun membranes with enhanced visible-light-induced photocatalytic performance. *J. Mater. Sci.* **2016**, *51*, 6974–6986. [CrossRef]
18. Li, X.; Wang, C.; Xia, N.; Jiang, M.; Liu, R.; Huang, J.; Li, Q.; Luo, Z.; Liu, L.; Xu, W.; et al. Novel ZnO-TiO2 nanocomposite arrays on Ti fabric for enhanced photocatalytic application. *J. Mol. Struct.* **2017**, *1148*, 347–355. [CrossRef]
19. Momeni, M.M.; Ghayeb, Y. Fabrication, characterization and photocatalytic properties of Au/TiO2-WO3 nanotubular composite synthesized by photo-assisted deposition and electrochemical anodizing methods. *J. Mol. Catal. A Chem.* **2016**, *417*, 107–115. [CrossRef]
20. Momeni, M.M. Fabrication of copper decorated tungsten oxide-titanium oxide nanotubes by photochemical deposition technique and their photocatalytic application under visible light. *Appl. Surf. Sci.* **2015**, *357*, 160–166. [CrossRef]
21. Momeni, M.M.; Ghayeb, Y. Cobalt modified tungsten–titania nanotube composite photoanodes for photoelectrochemical solar water splitting. *J. Mater. Sci. Mater. Electron.* **2016**, *27*, 3318–3327. [CrossRef]
22. Angkaew, S.; Limsuwan, P. Preparation of silver-titanium dioxide core-shell (Ag@TiO$_2$) nanoparticles: Effect of Ti-Ag mole ratio. *Procedia Eng.* **2012**, *32*, 649–655. [CrossRef]
23. Daskalaki, V.M.; Antoniadou, M.; Li Puma, G.; Kondarides, D.I.; Lianos, P. Solar light-responsive Pt/CdS/TiO$_2$ photocatalysts for hydrogen production and simultaneous degradation of inorganic or organic sacrificial agents in wastewater. *Environ. Sci. Technol.* **2010**, *44*, 7200–7205. [CrossRef]
24. Bian, Z.; Tachikawa, T.; Kim, W.; Choi, W.; Majima, T. Superior electron transport and photocatalytic abilities of metal-nanoparticle-loaded TiO$_2$ superstructures. *J. Phys. Chem. C* **2012**, *116*, 25444–25453. [CrossRef]
25. Tian, Y.; Tatsuma, T. Mechanisms and applications of plasmon-induced charge separation at TiO$_2$ films loaded with gold nanoparticles. *J. Am. Chem. Soc.* **2005**, *127*, 7632–7637. [CrossRef] [PubMed]
26. Seh, Z.W.; Liu, S.; Low, M.; Zhang, S.-Y.; Liu, Z.; Mlayah, A.; Han, M.-Y. Janus Au-TiO$_2$ photocatalysts with strong localization of plasmonic near-fields for efficient visible-light hydrogen generation. *Adv. Mater.* **2012**, *24*, 2310–2314. [CrossRef] [PubMed]
27. Al-Akraa, I.M.; Mohammad, A.M.; El-Deab, M.S.; El-Anadouli, B.E. Flower-shaped gold nanoparticles: Preparation, characterization, and electrocatalytic application. *Arab. J. Chem.* **2017**, *10*, 877–884. [CrossRef]
28. Santhoshkumar, J.; Rajeshkumar, S.; Venkat Kumar, S. Phyto-assisted synthesis, characterization and applications of gold nanoparticles—A review. *Biochem. Biophys. Rep.* **2017**, *11*, 46–57. [CrossRef] [PubMed]
29. Haruta, M. Catalysis of gold nanoparticles deposited on metal oxides. *Cattech* **2002**, *6*, 102–115. [CrossRef]
30. Thompson, D.T. Using gold nanoparticles for catalysis. *Nano Today* **2007**, *2*, 40–43. [CrossRef]
31. Barakat, T.; Rooke, J.C.; Genty, E.; Cousin, R.; Siffert, S.; Su, B.-L. Gold catalysts in environmental remediation and water-gas shift technologies. *Energy Environ. Sci.* **2013**, *6*, 371–391. [CrossRef]
32. Okumura, M.; Tanaka, K.; Ueda, A.; Haruta, M. The reactivities of dimethylgold(III)β-diketone on the surface of TiO2: A novel preparation method for Au catalysts. *Solid State Ion.* **1997**, *95*, 143–149. [CrossRef]
33. Su, R.; Tiruvalam, R.; He, Q.; Dimitratos, N.; Kesavan, L.; Hammond, C.; Lopez-Sanchez, J.A.; Bechstein, R.; Kiely, C.J.; Hutchings, G.J.; et al. Promotion of phenol photodecomposition over TiO$_2$ using Au, Pd, and Au–Pd nanoparticles. *ACS Nano* **2012**, *6*, 6284–6292. [CrossRef] [PubMed]
34. Haugen, A.B.; Kumakiri, I.; Simon, C.; Einarsrud, M.-A. TiO2, TiO2/Ag and TiO2/Au photocatalysts prepared by spray pyrolysis. *J. Eur. Ceram. Soc.* **2011**, *31*, 291–298. [CrossRef]
35. Chandrasekharan, N.; Kamat, P.V. Assembling gold nanoparticles as nanostructured films using an electrophoretic approach. *Nano Lett.* **2001**, *1*, 67–70. [CrossRef]
36. Fackler, J.P. Catalysis by Gold By Geoffrey C. Bond (Brunel University, U.K.), Catherine Louis (Université Pierre et Marie Curie, France), and David, T. Thompson (Consultant, World Gold Council, UK). From the Series: Catalytic Science Series, Volume 6. Series Edited by Graham J. Hutchings. Imperial College Press: London. 2006. xvi + 366 pp. ISBN 1-86094-658-5. *J. Am. Chem. Soc.* **2007**, *129*, 4107. [CrossRef]
37. Luna, M.; Gatica, J.M.; Vidal, H.; Mosquera, M.J. One-pot synthesis of Au/N-TiO$_2$ photocatalysts for environmental applications: Enhancement of dyes and NOx photodegradation. *Powder Technol.* **2019**, *355*, 793–807. [CrossRef]

38. Zhu, H.; Chen, X.; Zheng, Z.; Ke, X.; Jaatinen, E.; Zhao, J.; Guo, C.; Xie, T.; Wang, D. Mechanism of supported gold nanoparticles as photocatalysts under ultraviolet and visible light irradiation. *Chem. Commun.* **2009**, 7524–7526. [CrossRef]
39. Kamely, N.; Ujihara, M. Confeito-like Au/TiO$_2$ nanocomposite: Synthesis and plasmon-induced photocatalysis. *J. Nanoparticle Res.* **2018**, *20*, 172. [CrossRef]
40. Li, J.; Zeng, H.C. Preparation of monodisperse Au/TiO$_2$ nanocatalysts via self-assembly. *Chem. Mater.* **2006**, *18*, 4270–4277. [CrossRef]
41. D'Amato, C.A.; Giovannetti, R.; Zannotti, M.; Rommozzi, E.; Ferraro, S.; Seghetti, C.; Minicucci, M.; Gunnella, R.; Di Cicco, A. Enhancement of visible-light photoactivity by polypropylene coated plasmonic Au/TiO$_2$ for dye degradation in water solution. *Appl. Surf. Sci.* **2018**, *441*, 575–587. [CrossRef]
42. Singh, J.; Manna, A.K.; Soni, R.K. Bifunctional Au–TiO$_2$ thin films with enhanced photocatalytic activity and SERS based multiplexed detection of organic pollutant. *J. Mater. Sci. Mater. Electron.* **2019**, *30*, 16478–16493. [CrossRef]
43. Mompelat, S.; Le Bot, B.; Thomas, O. Occurrence and fate of pharmaceutical products and by-products, from resource to drinking water. *Environ. Int.* **2009**, *35*, 803–814. [CrossRef] [PubMed]
44. Pereira, A.M.P.T.; Silva, L.J.G.; Meisel, L.M.; Lino, C.M.; Pena, A. Environmental impact of pharmaceuticals from Portuguese wastewaters: Geographical and seasonal occurrence, removal and risk assessment. *Environ. Res.* **2015**, *136*, 108–119. [CrossRef] [PubMed]
45. Song, M.; Bian, L.; Zhou, T.; Zhao, X. Surface ζ potential and photocatalytic activity of rare earths doped TiO$_2$. *J. Rare Earths* **2008**, *26*, 693–699. [CrossRef]
46. Sentein, C.; Guizard, B.; Giraud, S.; Yé, C.; Ténégal, F. Dispersion and stability of TiO2nanoparticles synthesized by laser pyrolysis in aqueous suspensions. *J. Phys. Conf. Ser.* **2009**, *170*, 012013. [CrossRef]
47. Chen, W.; Zhang, J.; Cai, W. Sonochemical preparation of Au, Ag, Pd/SiO$_2$ mesoporous nanocomposites. *Scr. Mater.* **2003**, *48*, 1061–1066. [CrossRef]
48. Kuge, K.i.; Calzaferri, G. Gold-loaded zeolite A. *Microporous Mesoporous Mater.* **2003**, *66*, 15–20. [CrossRef]
49. Mihai, S.; Cursaru, D.; Ghita, D.; Dinescu, A. Morpho ierarhic TiO$_2$ with plasmonic gold decoration for highly active photocatalysis properties. *Mater. Lett.* **2016**, *162*, 222–225. [CrossRef]
50. Cojocaru, B.; Andrei, V.; Tudorache, M.; Lin, F.; Cadigan, C.; Richards, R.; Parvulescu, V.I. Enhanced photo-degradation of bisphenol pollutants onto gold-modified photocatalysts. *Catal. Today* **2017**, *284*, 153–159. [CrossRef]
51. Kumar, S.G.; Devi, L.G. Review on modified TiO2 photocatalysis under UV/visible light: Selected results and related mechanisms on interfacial charge carrier transfer dynamics. *J. Phys. Chem. A* **2011**, *115*, 13211–13241. [CrossRef]
52. Lu, M. *Photocatalysis and Water Purification: From Fundamentals to Recent Applications*; John Wiley & Sons: Hoboken, NJ, USA, 2013.
53. Wang, H.; Faria, J.L.; Dong, S.; Chang, Y. Mesoporous Au/TiO$_2$ composites preparation, characterization, and photocatalytic properties. *Mater. Sci. Eng. B* **2012**, *177*, 913–919. [CrossRef]
54. Wang, X.; Caruso, R.A. Enhancing photocatalytic activity of titania materials by using porous structures and the addition of gold nanoparticles. *J. Mater. Chem.* **2011**, *21*, 20–28. [CrossRef]
55. Wongwisate, P.; Chavadej, S.; Gulari, E.; Sreethawong, T.; Rangsunvigit, P. Effects of monometallic and bimetallic Au–Ag supported on sol–gel TiO2 on photocatalytic degradation of 4-chlorophenol and its intermediates. *Desalination* **2011**, *272*, 154–163. [CrossRef]
56. Scarabelli, L.; Sánchez-Iglesias, A.; Pérez-Juste, J.; Liz-Marzán, L.M. A "tips and tricks" practical guide to the synthesis of gold nanorods. *J. Phys. Chem. Lett.* **2015**, *6*, 4270–4279. [CrossRef] [PubMed]
57. Emeline, A.V.; Ryabchuk, V.; Serpone, N. Factors affecting the efficiency of a photocatalyzed process in aqueous metal-oxide dispersions: Prospect of distinguishing between two kinetic models. *J. Photochem. Photobiol. A Chem.* **2000**, *133*, 89–97. [CrossRef]
58. Evgenidou, E.; Fytianos, K.; Poulios, I. Semiconductor-sensitized photodegradation of dichlorvos in water using TiO$_2$ and ZnO as catalysts. *Appl. Catal. B Environ.* **2005**, *59*, 81–89. [CrossRef]
59. Le, H.N.; Babick, F.; Kühn, K.; Nguyen, M.T.; Stintz, M.; Cuniberti, G. Impact of ultrasonic dispersion on the photocatalytic activity of titania aggregates. *Beilstein J. Nanotechnol.* **2015**, *6*, 2423–2430. [CrossRef]
60. Behnajady, M.A.; Modirshahla, N.; Hamzavi, R. Kinetic study on photocatalytic degradation of C.I. Acid Yellow 23 by ZnO photocatalyst. *J. Hazard. Mater.* **2006**, *133*, 226–232. [CrossRef]

61. Grimme, S.; Bannwarth, C.; Shushkov, P. A robust and accurate tight-binding quantum chemical method for structures, vibrational frequencies, and noncovalent interactions of large molecular systems parametrized for all spd-block elements (Z = 1–86). *J. Chem. Theory Comput.* **2017**, *13*, 1989–2009. [CrossRef]
62. Lamiel-Garcia, O.; Ko, K.C.; Lee, J.Y.; Bromley, S.T.; Illas, F. When anatase nanoparticles become bulklike: Properties of realistic TiO$_2$ nanoparticles in the 1–6 nm size range from all electron relativistic density functional theory based calculations. *J. Chem. Theory Comput.* **2017**, *13*, 1785–1793. [CrossRef]
63. Zhang, X.; Sun, C.Q.; Hirao, H. Guanine binding to gold nanoparticles through nonbonding interactions. *Phys. Chem. Chem. Phys.* **2013**, *15*, 19284–19292. [CrossRef]
64. Marenich, A.V.; Jerome, S.V.; Cramer, C.J.; Truhlar, D.G. Charge model 5: An extension of hirshfeld population analysis for the accurate description of molecular interactions in gaseous and condensed phases. *J. Chem. Theory Comput.* **2012**, *8*, 527–541. [CrossRef] [PubMed]
65. Green, I.X.; Tang, W.; Neurock, M.; Yates, J.T. Insights into catalytic oxidation at the Au/TiO$_2$ dual perimeter sites. *Acc. Chem. Res.* **2014**, *47*, 805–815. [CrossRef] [PubMed]
66. Sakthivel, S.; Hidalgo, M.C.; Bahnemann, D.W.; Geissen, S.U.; Murugesan, V.; Vogelpohl, A. A fine route to tune the photocatalytic activity of TiO$_2$. *Appl. Catal. B Environ.* **2006**, *63*, 31–40. [CrossRef]

© 2020 by the authors. Licensee MDPI, Basel, Switzerland. This article is an open access article distributed under the terms and conditions of the Creative Commons Attribution (CC BY) license (http://creativecommons.org/licenses/by/4.0/).

Article

TiO₂ and Active Coated Glass Photodegradation of Ibuprofen

Samer Khalaf [1,2,*], **Jawad H. Shoqeir** [1], **Filomena Lelario** [2], **Sabino A. Bufo** [2], **Rafik Karaman** [2,3] **and Laura Scrano** [4]

[1] Soil & Hydrology Research Lab (SHR), Department of Earth and Environmental Sciences, Al-Quds University, 20002 Jerusalem, Palestine; jhassan@staff.alquds.edu
[2] Department of Sciences, University of Basilicata, Viadell'AteneoLucano 10, 85100 Potenza, Italy; filomenalelario@hotmail.com (F.L.); sabino.bufo@unibas.it (S.A.B.); dr_karaman@yahoo.com (R.K.)
[3] Department of Bioorganic Chemistry, Faculty of Pharmacy, Al-Quds University, 20002 Jerusalem, Palestine
[4] Department of European Cultures (DICEM), University of Basilicata, Via dell'AteneoLucano10, 85100 Potenza, Italy; laura.scrano@unibas.it
* Correspondence: skhalaf@staff.alquds.edu; Tel.: +970-598-600-781

Received: 29 January 2020; Accepted: 25 February 2020; Published: 18 May 2020

Abstract: Commercial non-steroidal anti-inflammatory drugs (NSAIDs) are considered as toxic to the environment since they induce side effects when consumed by humans or aquatic life. Ibuprofen is a member of the NSAID family and is widely used as an anti-inflammatory and painkiller agent. Photolysis is a potentially important method of degradation for several emerging contaminants, and individual compounds can undergo photolysis to various degrees, depending on their chemical structure. The efficiency oftitanium dioxide (TiO_2) and photocatalysis was investigated for the removal of ibuprofen from the aquatic environment, and the performance of these different processes was evaluated. In heterogeneous photocatalysis, two experiments were carried out using TiO_2 as (i) dispersed powder, and (ii) TiO_2 immobilized on the active surface of commercial coated glass. The kinetics of each photoreaction was determined, and the identification of the photoproducts was carried out by liquid chromatography coupled with Fourier-transform ion cyclotron resonance mass spectrometry (LC-FTICR MS). The overall results suggest that the TiO_2 active thin layer immobilized on the glass substrate can avoid recovery problems related to the use of TiO_2 powder in heterogeneous photocatalysis and may be a promising tool toward protecting the environment from emerging contaminants such as ibuprofen and its derivatives.

Keywords: ibuprofen; advanced oxidation process; TiO_2; photocatalysis; active glass

1. Introduction

Emerging contaminants resulting from the presence and circulation of pharmaceuticals (PhCs) were the focus of many environmental chemists over the last few decades. In the aquatic environment, PhCs are introduced anthropogenically through pharmaceutical or conventional plants [1]. PhCsare found in tiny concentrations in surface waters, indicating insufficient treatment of such entities during the standard sewage treatment processes (STPs). The occurrence of these toxic drugsin wastewater effluent, along with their metabolites which may be much more harmful than their parent compounds [2,3], has the potential to be a great health problem since they are endocrine-disrupting agents, thus posing a significant barrier to the use of water recycling [4].

Ibuprofen (IBP), (*RS*)-2-(4-(2-methylpropyl) phenyl) propanoic acid, shown in Figure 1, is a non-steroidal anti-inflammatory drug(NSAID) belonging to the class of propanoic acid derivatives used as pain relief for several inflammation conditions, including rheumatoid arthritis, asan analgesic for pain relief in general, and as an antipyretic to help in fever conditions [5]. IBP enters the aquatic environment through effluents exiting secondary wastewater treatment plants, which are inefficient in removing a variety of small organic molecules, particularly pharmaceuticals [6]. Reported studies showed that the concentrations of IBP found in rivers and other environmental waters range between 10 ng·L^{-1} and 169 µg·L^{-1} [7].

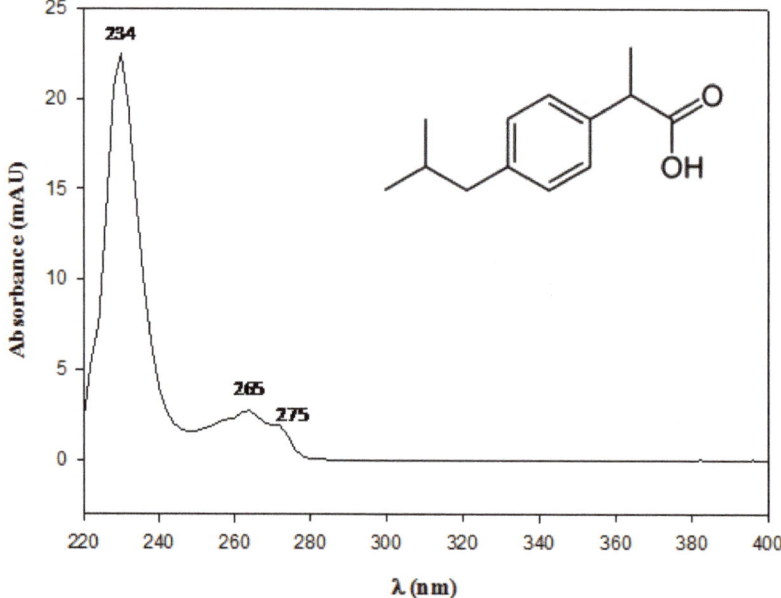

Figure 1. Chemical structure and ultraviolet (UV) absorbance of ibuprofen (IBP).

The fact that the current conventional wastewater treatment technologies such as those based on biological, thermal, and physical treatment processes are not efficient in removing or degrading small-molecular-weight pharmaceuticals with low biodegradability and high chemical stability such as ibuprofen [8] encouraged us to devote considerable effort toward developing a novel purification method that can efficiently remove this recalcitrant organic contaminant from the water environment.

Recently, we found that the integration of separation technologies, consisting of sequential elements of ultra-filtration (UF), activated carbon filtration (AC), and reverse osmosis (RO), as well as adsorption technology based on a surface modified clay minerals, was efficient in removing IBP and other pharmaceuticals to a safe level [9–13]. Nevertheless, the operating principles of these tools are only based on phase-transfer technologies, whereby the contaminant is retained on the filter or adsorbent without being degraded or destroyed to non-toxic compounds. Furthermore, some of the technologies used, such as UF and RO, are too expensive to be adopted in most real environmental situations. For the abovementioned reasons, we successfully attempted to find a good alternative method (degradation via photocatalysis) to these technologies for removing such pollutants from the aquatic environment.

The growing awareness of the risk arising from the occurrence of toxic organic contaminants in the aquatic environment promoted the development of technologies, such as photodegradation, and other advanced oxidation processes (AOPs), for efficient destruction of organic toxic compounds that exist in water and wastewater, including PhCs [14–16].

AOPs, i.e., processes based on highly reactive species such as hydroxyl radical (•OH), can oxidize and mineralize practically every organic entity [3], yielding CO_2 and inorganic ions, thus resulting in total destruction of the target pollutant [3,8,16]. Advanced oxidation processes (AOPs) involve several homogeneous and heterogeneous processes such as photolysis, photocatalysis, ozonation, electrochemical oxidation, photo-Fenton, wet air oxidation, and sonolysis [8,15,16]. The most popular and effective type of AOP employed in water and wastewater treatment is heterogeneous photocatalysis with semiconductors [17–19].

Heterogeneous photocatalysis is a process via which a photoreaction is accelerated by the presence of a catalyst (usually semiconductor). In order for this to occur, the dispersed solid particles of the semiconductor in the treated solutionshould absorb significant portions of the UV light, and, under radiation, they may be photo-excited and produce oxidizing agents from water and oxygen [19]. Generally, TiO_2 is considered the most efficient semiconductor to be employed in photocatalysis because of several factors including its low cost, low toxicity, chemical stability, large band gap, and high photosensitivity [8,16–19].

Under ultraviolet irradiation, TiO_2 as a semiconductor causes the jump of an electron (e−) from the valence band (VB) to the conduction band (CB), resulting in the formation of a positive hole (h+) at the site of the electron. In the presence of aqueous suspended TiO_2, the hole and electron can produce radicals of hydroxyl and superoxide that are very potent in the oxidation of many kinds of organic entities found in water sources, thereby leading to total degradation of these toxic organic agents. [20,21]. Equations (1)–(3) depict the formation reactions of the superoxide and hydroxyl radicals upon catalysis with TiO_2

$$TiO_2 + h\nu \rightarrow e_{CB-} + h_{VB+}. \quad (1)$$

$$H_2O + h_{VB+} \rightarrow OH^\bullet + H_{aq+}. \quad (2)$$

$$O_{2(ads)} + e_{CB-} \rightarrow O_2^\bullet\text{-}(ads) + H_2O + OH^\bullet. \quad (3)$$

In some processes, a complete degradation of organic pollutants requires the presence of a radiation source, oxidizing agent, and a semiconductor. Oxidation and reduction processes are promoted by photo-generated charge carriers resulting from the excitation of TiO_2 via photons with higher energy. Currently, this kind of photocatalysis is utilized to purify water [8,16–19].

Although photocatalytic degradation using suspensions of TiO_2 particles was extensively employed to catalyze different contaminants, such as drugs, and although it achieved good results in recent years, this technique fails to be widely used because of the high cost and difficulty in isolating the semiconductor from the mixture after degradation [8,16–19]. To find a way around the need for catalyst recovery via filtration, a different approach, consisting of catalyst immobilization on a stationary support, should be assessed. For fulfilling this aim, TiO_2 immobilized on different materials instead of the traditional powder was advocated and tested for obtaining a promising clean treatment method with a low cost [22–24].

In this study, the efficiency of two different systems, direct photolysis and heterogeneous photocatalysis (TiO_2 powder and TiO_2 immobilized on active glass), was investigated using simulated solar irradiation for the removal of IBP and its major photoproducts from the aqueous phase. The comparative performance of the adopted process was analyzed under the same experimental conditions, and the kinetics for each photodegradation reaction was evaluated. Moreover, major photoproducts were detected and identified via liquid chromatography coupled with Fourier-transform ion cyclotron resonance mass spectrometry (LC-FTICR MS).

It should be emphasized that the study on the photodegradation of IBP using an immobilized TiO_2 system can be regarded as representative of a process for the degradation of a variety of pollutants which impose a risk to the environment.

2. Results and Discussion

2.1. Characterization of Active TiO_2-Coated Glass

Figure 2A,B depict the SEM image of the active glass and the TiO_2 coating comb geometry on the glass surface, respectively. The TiO_2 film thickness was 397.2 nm, as shown in Figure 2A.

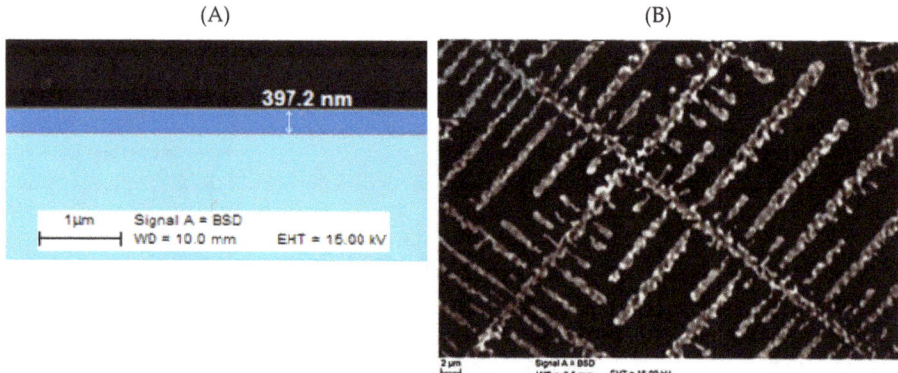

Figure 2. (**A**) SEM image of the blue glass cross-section illustrating the position of the TiO$_2$ layer immobilized on the glass surface; (**B**) SEM image of the blue glass surface illustrating the fine-tooth comb nature of the TiO$_2$ coating.

Table 1 lists the elements in the glass surface and core class. As shown in the table, TiO$_2$ is present only in the glass surface along with other elements, whereas it is absent in the core glass.

Table 1. Scanning electron microscopy EDX analysis of the glass surface coated with TiO$_2$ as compared to the glass core composition.

Elemental Composition of the Core Glass Energy: 6.686 KeV, Counts: 250		Elemental Composition of the Grafted Surface Energy: 6.686 KeV, Counts: 250	
Wt.% ± SD	Elements	Wt.% ± SD	Elements
47.2 ± 0.4	O	44.4 ± 0.4	O
36.2 ± 0.2	Si	38.6 ± 0.2	Si
4.3 ± 0.1	Na	6.0 ± 0.2	Na
6.0 ± 0.1	Ca	5.5 ± 0.2	Ca
2.4 ± 0.1	Mg	2.4 ± 0.1	Mg
2.1 ± 0.1	Sn	1.5 ± 0.1	Ti
0.3 ± 0.05	K	0.3 ± 0.05	K
1.0 ± 0.05	Al	0.3 ± 0.05	Al
0.4 ± 0.1	Fe	0.3 ± 0.05	Fe
0.1 ± 0.05	Cl	-	-

Sheel et al. (1998) reported the presence of cobalt oxide in a concentration of less than 75 µg/g [24], whereas we did not succeed in detecting any amount of cobalt oxide using our technique. It is believed that the blue color of the glass is due to the presence of cobalt oxide. For this reason, we report here the TiO$_2$-coated active glass using the abbreviation "blue glass".

The glass coating is based on photocatalytic anatase TiO$_2$, which is the most effective known photocatalyst [22,23]. For the preparation of the glass coating, TiO$_2$ nanocrystallinefilm is deposed onto float glass using an atmospheric pressure chemical vapor deposition technique (APCVD) as described in Reference [24]. Anatase is considered a more efficient photocatalyst than rutile because of its slower rate of recombination [23,24]. In the coated surface of the active glass, the catalyzing efficiency is improved by the presence of Fe$_3$O$_4$, which favors the formation of multiple band gaps, enlarging the wavelength range that can be absorbed by the glass surface [22,23].

2.2. Photodegradation Experiments

2.2.1. Preliminary Study

Prior to photodegradation of IBP, which involves photolysis and photocatalysis experiments, thermal (45 °C) or hydrolytic reactions in pure IBP solution and the adsorption of the pharmaceutical on the catalyst surface in TiO_2 powder suspension were assessed. No significant loss of IBP occurred in dark conditions due to thermal reactions or hydrolysis. The adsorption equilibrium on TiO_2 powder was reached within 60 min, and a slight decrease (not more than 4.3%) of free IBP concentration in the 0.2–0.3 g·L^{-1} TiO_2 suspensions was achieved. Adsorption increased to 6.9% and more (7.8%) when the amount of TiO_2 powder was augmented to 0.4–0.5 g·L^{-1}. The mentioned values are the average of three replicates.

2.2.2. Photolysis Experiment

The uranyl oxalate method [25,26] was used to assess the light emission effectiveness of the irradiation system prior to the experimental work. The disappearance of oxalate was 7.2×10^{-4} mol·s^{-1}.

Despite the high efficiency of the irradiation system (irradiance to exposed surface of the reactor, 500 W/m^2), IBP concentration during this experiment was decreased only by 8.4% after more than 20 h. This result indicates that IBP is stable during photolysis because it has great chemical stability and a reduced molar adsorption coefficient above 280 nm [19] (Figure 1).

2.2.3. Photocatalysis Experiments

TiO_2 commercial powder and TiO_2-coated active glass were employed separately. The degradation efficiency via the two different methods was compared.

Photocatalytic Degradation Using TiO_2 Powder

The concentration of IBP during the photocatalysis reaction was monitored using High Performance Liquid Chromatography-Ultra Violet HPLC-UV. The standard solution used showed a peak of IBP at 4.1 min retention time. After 5 min of sample irradiation, a marked decrease in the IBP concentration was observed (21.8% of the initial concentration) along with the appearance of a new photoproduct (Figure 3A,B).

After 30 min, a new peak due to the formation of another derivative was observed, and an approximately 60% reduction in IBP initial concentration was obtained (Figure 3C). Complete disappearance of IBP was achieved after 270 min (Figure 3D), while complete depletion including derivatives was observed after approximately 23 hours (Figure 3E).

The combined results demonstrate that IBP was completely degraded by photocatalytic oxidation using TiO_2 powder under simulated solar irradiation, and the efficiency of its removal was more than 87% within 80 min. The degradation of IBP occurred as a result of a photo-irradiation of the semiconductor, causing an electron transfer to the conduction band which subsequently formed a hole in the valence band, which led to photo-induced charge separation on the semiconductor surface and an exchange of electrons on the water semiconductor interface. This led to the formation of $^\bullet O_2^-$ via interactions of adsorbed oxygen molecules with the photo-generated conduction band electrons, whereas the $^\bullet OH$ generated from the oxidation of the adsorbed water or hydroxyl anions by the valence band hole oxidized the adsorbed IBP molecules [17,19].

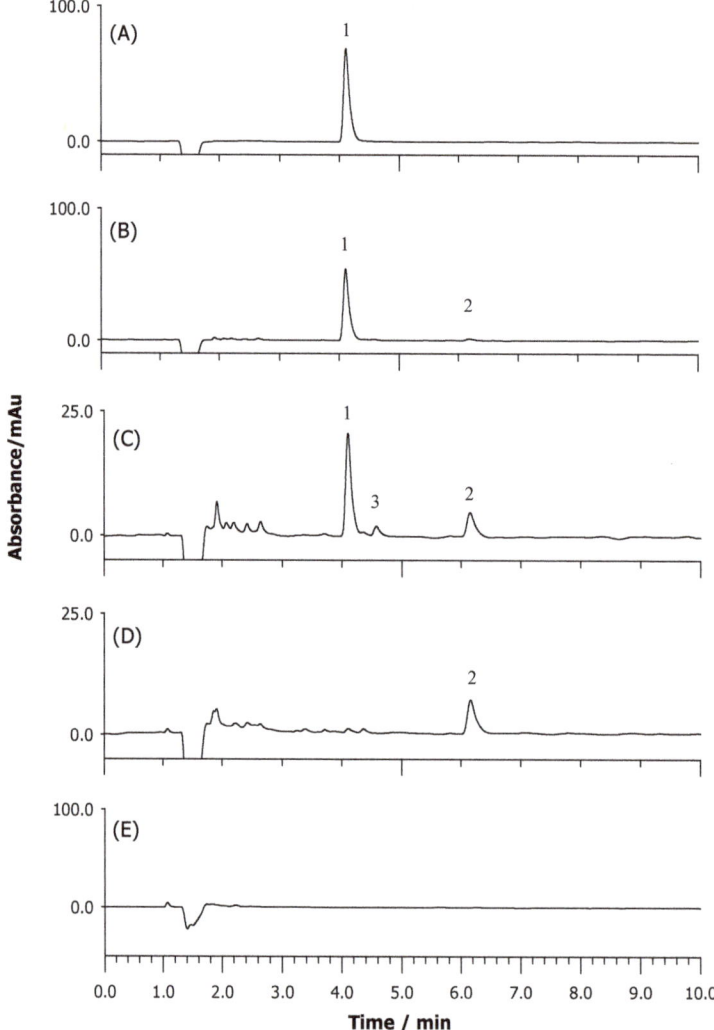

Figure 3. HPLC-UV separation of photodegraded solution using TiO_2 powder, IBP (1), and IBP photoproducts (2 and 3) (**A**) at time zero (initial standard solution), (**B**) after 5 min, (**C**) after 30 min, (**D**) after 270 min with complete disappearance of IBP, and (**E**) after 23 h with complete disappearance of IBP photoproducts.

Photocatalytic Degradation Using TiO_2-Coated Active Glass

During the first 2 h, a slight decrease in IBP concentration (2.7%) was achieved, while the appearance of a unique photoproduct was accomplished after 3 h (Figure 4A,B). Following a further 9 h of irradiation, the concentration of IBP was decreased to 50%, and, after 24 h, IBP and its photoproduct disappeared (Figure 4C).

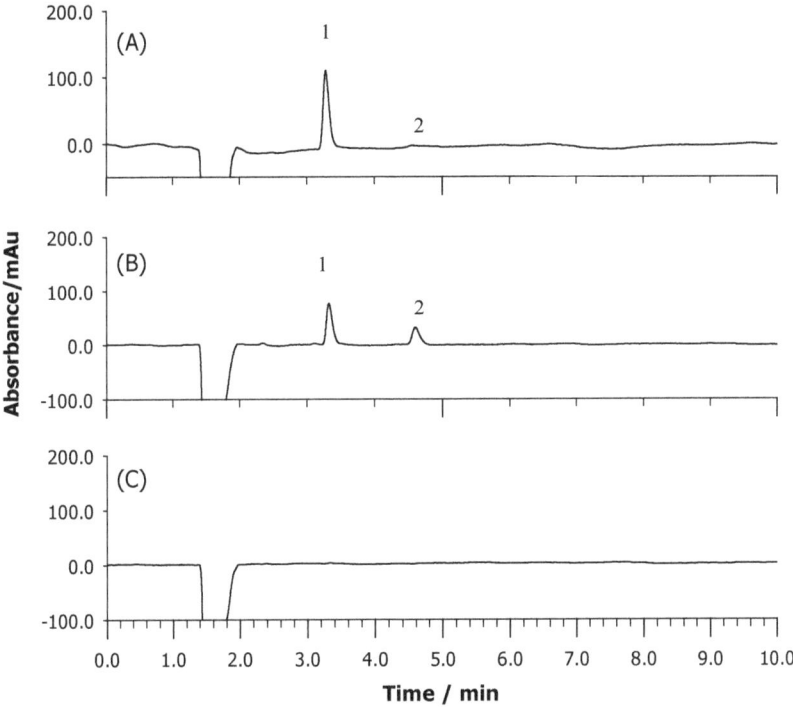

Figure 4. HPLC-UV separation of photo-degraded solution using TiO_2-coated blue glass, IBP (1), and IBP photoproducts (2) (**A**) after 180 min, (**B**) after 10 h, and (**C**) after 24 h with complete disappearance of IBP and IBP photoproducts.

2.3. Kinetics Studies

2.3.1. Experimental Observations

No degradation of IBP was observed in the dark in all aqueous environments adopted for the experiments. Direct photolysis under simulated sunlight did not achieve the desired goal. Accordingly, we can conclude that the direct interaction of IBP with sunlight (both via thermal hydrolytic reactions and photolysis) cannot lead to IBP's quick degradation. However, in the presence of TiO_2, complete removal of this NSAID was obtained although a xenon lamp with low UV energy was used for irradiation aiming at the simulation of sunlight effect.

Different amounts (0.1–0.5 g·L^{-1}) of TiO_2 micro-particles were added to a solution of 25 mg·L^{-1} IBP to determine the efficiency of the catalytic process. The half-life (experimentally observed) of the mother molecule was reduced upon increasing the concentration of the catalyst from 0.1 to 0.2 g·L^{-1}, and it remained constant upon adding an amount of 0.3 g·L^{-1}, while it increased when concentrations of 0.4 or 0.5 g·L^{-1} were tested. The rationale behind such behavior is that the number of IBP molecules and photons absorbed on the TiO_2 particles increased with a moderate increase in the catalyst loading; however, with further addition of the semiconductor (powder), the phenomenon of light scattering took place and the number of useful photons per mass unit of TiO_2 was reduced. The disappearance of IBP at the highest concentrations of TiO_2 powder was mostly due to its mere physical adsorption onto the surface of semiconductor particles.

Figure 5A,B illustrates the depletion trend of IBP measured as C_t/C_0 versus irradiation time (A) and evolution of photoproducts (B) using different photodegradation methods. In both photocatalysis processes, IBP underwent complete disappearance via the formation of one or two intermediates that were subsequently removed within 24 h.

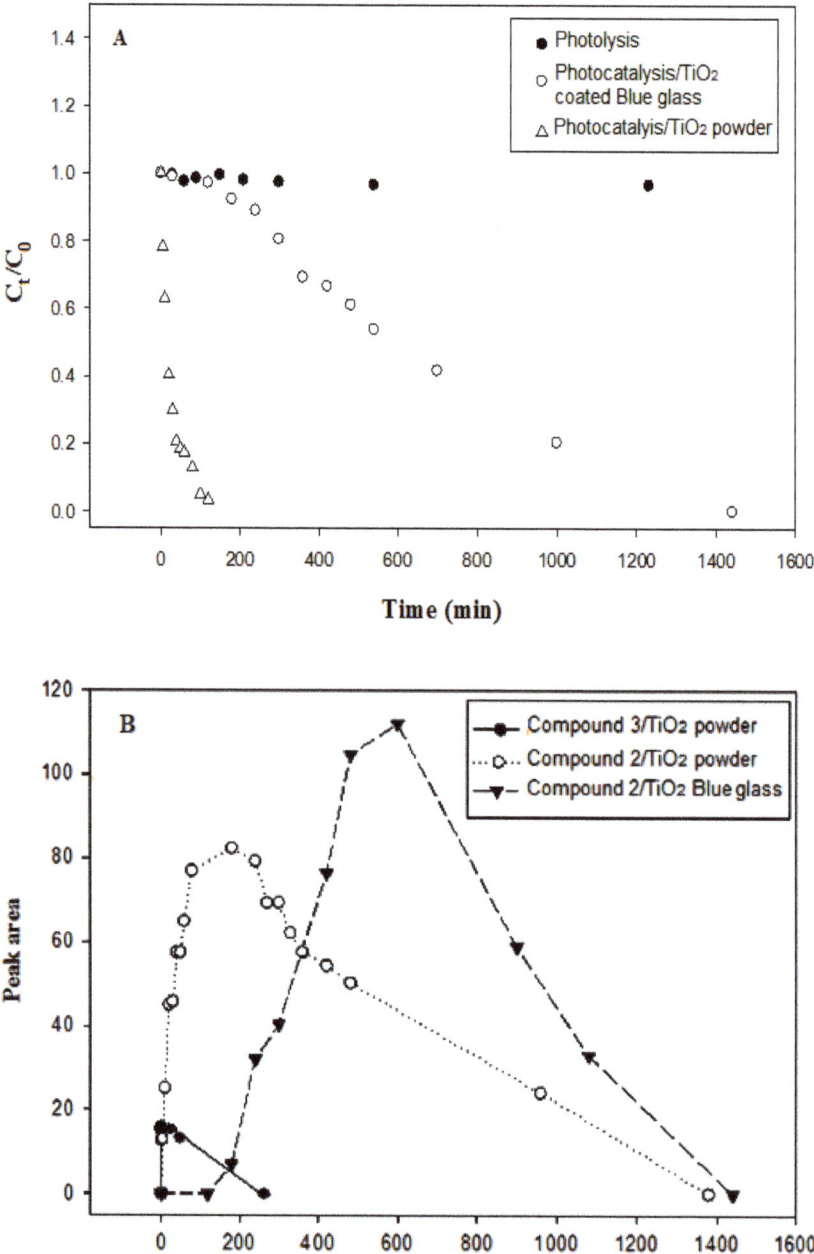

Figure 5. Evaluation of IBP degradation measured as C_t/C_0 versus irradiation time (**A**); evolution of photoproducts using different photodegradation methods (**B**).

In the photocatalysis experiment with TiO_2-coated active glass, the reaction was apparently slower than degradation obtained using TiO_2 powder, but a satisfactory depletion of IBP and its derivatives was reached in approximately the same time.

2.3.2. Kinetic Parameters

To find the kinetics model, kinetic parameters were calculated using integrated equations describing zero-, first-, and second-order (Langmuir-Hinshelwood) order equations [27]. According to Snedecor and Cochran (1989) [28], the least square method should be utilized to find the best fit.

Table 2 summarizes the kinetic parameters of IBP degradation under the photocatalysis experiment with TiO_2 powder.

Table 2. Kinetic parameters of IBP degradation under photocatalysis experiments.

Catalyzer	Reaction Order	Linearized Rate Equation	$(\Sigma LSq)/n$	k	$t_{1/2}$
TiO_2 powder	Zeroorder	$C_t = C_0 - kt$	12.47	0.2378 mg·L^{-1}·min^{-1}	37.6 min
	Firstorder	$\ln C_t = \ln C_0 - kt$	6.24	0.0251 min^{-1}	27.6 min
	Secondorder	$C_0/C_t = 1 + (1/t_{1/2})t$	0.18	0.0034 L·mg^{-1}·min^{-1}	11.8 min
TiO_2-coated active glass	Zeroorder	$C_t = C_0 - kt$	1.82	0.0124 mg·L^{-1}·min^{-1}	785 min
	Firstorder	$\ln C_t = \ln C_0 - kt$	0.05	0.0012 min^{-1}	575 min
	Secondorder	$C_0/C_t = 1 + (1/t_{1/2})t$	8.62	0.0001 L·mg^{-1}·min^{-1}	320 min

ΣLSq, sum of least squares = $\Sigma n (Cexp-Ccalc)^2$; Cexp, experimental values of concentrations; Ccalc, value of concentrations calculated from rate equations; n, number of experimental observations; k, kinetic constant; $t_{1/2}$, half-life.

It must be taken into account that, owing to the dissimilar units associated with them, the values of kinetic constants calculated by equations describing reactions of different order cannot be compared. For this reason, it is useful to consider the values of half-life, which are always expressed in time units. Table 2 shows dissimilar values of the half-life when calculated using different equations applied to the same system. The least square method of estimation is a powerful method to assess the equation that can best fit the experimental data.

Apparently, the measured reaction rate of IBP under irradiation conditions using TiO_2 powder as a catalyst was best fit by a Langmuir–Hinshelwood-type equation [29].

$$C_t = C_0 \, t_{1/2}/(t + t_{1/2}), \tag{4}$$

where C_0 is the initial amount (mg) of IBP per liter of solution, C_t is the remaining concentration at time t, and $t_{1/2}$ is the half-life of the reactant.

Equation (4) shows the minimum value of the sum of least squares, based on the number of observations $(\Sigma LSq)/n$, and describes a second-order reaction governed by the kinetic law.

$$v = -dC_t/dt = kC_t^2, \tag{5}$$

where v is the reaction rate, and k is the rate (or kinetic) constant [27,29], which in our case can be calculated as

$$k = 1/(C_0 t_{1/2}). \tag{6}$$

Equation (5) represents a double dependence of the reaction velocity on the concentration.

The rationale behind such a finding may be due to the total reaction rate during the photocatalytic process being affected by two sorption states, both depending on the dissolved concentration of the pharmaceutical. The amount of reactant disappearing at each time t is affected by its free concentration in the powder suspension, as well as by the amount adsorbed on the catalyst particles, which depends on the remaining free concentration of IBP.

The half-life value for a second-order reaction, calculated by means of the linearized form of Equation (4) (Table 2), was just 11.8 min, while, after 80 min, 87% of IBP was converted.

The second-order kinetics shown in Figure 6A was confirmed by the linear behavior of (C_0/C_t) as a function of irradiation time (Figure 6B).

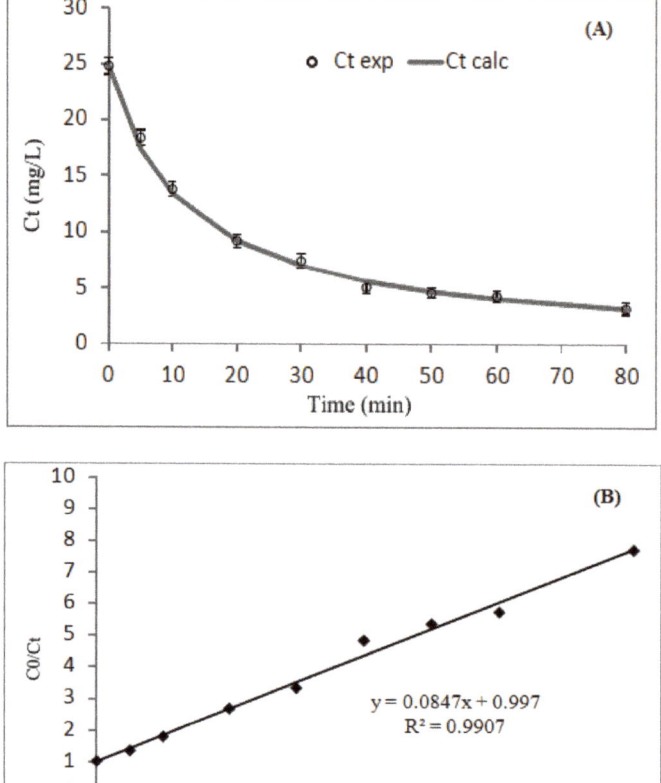

Figure 6. (**A**) Photodegradation of ibuprofen catalyzed by TiO_2 powder; Ct calc, values calculated using Equation (4); Ct exp, experimental values; error bars represent the standard deviations of three replicate experiments. (**B**) Trend of second-order linearized equation used for the calculation of kinetic parameters reported in Table 2.

From the results, it can be remarked that the initial degradation rate was high; however, it decreased rapidly as the reaction proceeded. The degradation was fast during the first 20 min, and then it gradually decreased; this trend is typical of second-order reactions. Several observations can be related to such a behavior: (1) the high concentration of IBP at the beginning of the reaction facilitates the useful attack by the hydroxyl radicals, resulting in high degradation rate; however, when IBP concentration gradually decreases, the degradation rate subsequently decelerates due to the dilution effect that reduces the possibility of useful collisions with the hydroxyl radicals; (2) the competitive reactions of the hydroxyl radicals with IBP degradation products that are produced during the reaction; (3) the recombination reactions of radical–radical.

The photodegradation reaction of IBP catalyzed by TiO_2 immobilized on active glass surface achieved 85% compound disappearance after 24 h of simulated solar light irradiation. By attempting to fit the concentration values vs. time using various-order integrated kinetic equations, it was found that the data best fit the first-order kinetic equation

$$C_t = C_0 e^{-kt}. \tag{7}$$

In Figure 7A, the best fit of experimental data calculated using Equation (7) is represented clearly as confirmed by the high value of the determination coefficient (R^2) obtained for the linearized form of Equation (7) (Table 2, Figure 7B).

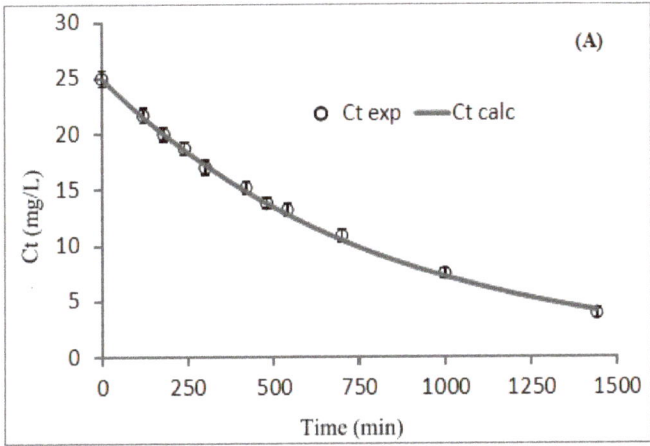

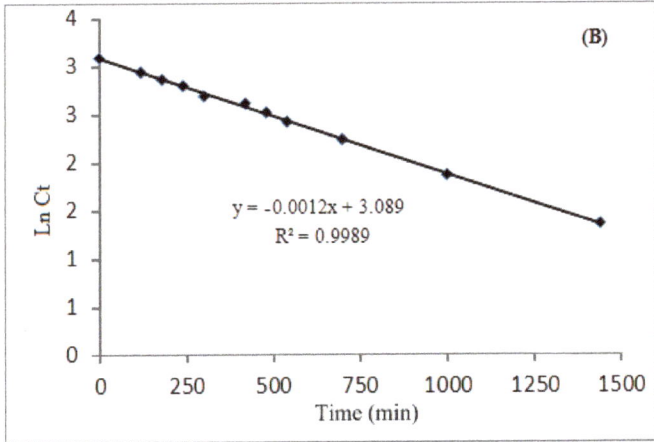

Figure 7. (**A**) Photodegradation of ibuprofen catalyzed by TiO_2-coated blue glass; $C_t$calc, values calculated using Equation (6); Ct exp, experimental values; error bars represent the standard deviations of three replicate experiments. (**B**) Trend of first-order linearized equation used for the calculation of kinetic parameters reported in Table 2.

In this case, the half-life can be calculated as

$$t_{1/2} = Ln2/k. \tag{8}$$

The value obtained was 575 min (Table 2), which is far from the half-life resulting from the experiment with TiO$_2$ powder; nevertheless, it is satisfactory if we consider the high stability shown by IBP molecules not only in the darkness (for testing thermal and hydrolysis degradation), but also under light irradiation (photolysis degradation). Moreover, from Equation (8), it is possible to notice that, unlike the case of the second-order reaction, the half-life for the first-order reaction does not depend on the initial concentration of the reactant. This means that the reaction catalyzed by TiO$_2$ immobilized on the active glass surface does not suffer from the same limitations encountered in the case of the second-order kinetics shown by IBP under the photoreaction catalyzed by TiO$_2$ powder dispersion (light scattering, radical–radical recombination reactions, and dilution effect). Furthermore, it should be emphasized that (i) the number and persistence of derivatives was reduced in the case of coated active glass, and (ii) the time needed for an efficient degradation of the mother drug and its derivatives was approximately the same.

We have to mention that the degradation of IBP is also influenced by the pH value of the medium [30]. The production of hydroxyl radicals is generally increased in an alkaline medium, since high concentrations of OH$^-$ result in the formation of hydroxyl radicals, which are produced from the reaction of OH$^-$ with TiO$_2$ on its surface's holes [31]. The pH value can also affect the charge on the catalyst particles; consequently, the electrostatic interactions between the charged surface of TiO$_2$ and the pollutant molecules can be largely influenced, thus leading to a change in the adsorption level of these molecules on the catalyst surface and interfacial electron transfer [32]. The most dominant factor affecting the adsorption of pollutant on catalyst surface is the catalyst zero-point charge (zpc), which is defined as the pH at which the surface of the catalyst has neutral charge [33].

For TiO$_2$ P25 Evonik-Degussa, the zero-point charge value is 6.9. Therefore, the surface of TiO$_2$ is positively charged in acidic media and negatively charged in basic media [33]. Accordingly, the effect of pH depends mainly on the type of the pollutant and the zero = point charge of the semiconductor. As IBP is weakly acidic in nature, it is expected to be negatively charged at pH higher than 3 [30], while the TiO$_2$ surface is positively charged at pH less than 6.9 [34]. Therefore, at pH = 4.5 where the photocatalytic experiment took place, the adsorption of IBP and, consequently, its photocatalytic oxidation were favored [30].

2.4. Identification of Intermediate Photoproducts

For the identification of IBP degradation byproducts, samples were collected at various time intervals and analyzed and identified by LC-FTICR MS system in the *m/z* range of 50–1000 in negative ionization mode. The results indicate the formation of two major photoproducts (Table 3). In addition, they reveal that the hydroxyl radicals attacked both the propionic acid and isobutyl substituent in IBP, resulting in the formation of two products, 2-[4-(1-hydroxyisobutyl) phenyl] propionic acid (2) and 4-(1-hydroxy isobutyl) acetophenone (3). Figure 8 depicts the proposed reaction pathways. The peak that appears at a nominal *m/z* value of 221, showing the formation of a mono-hydroxylated product of IBP, corresponds to 1-hydroxy IBP (2). Furthermore, product 2 was converted into another derivative with a nominal *m/z* value of 191, which corresponds to 4-(1-hydroxy isobutyl) acetophenone (3).

Table 3. Identification of ibuprofen and its photoproducts during photocatalytic degradation as deprotonated molecules, [M–H]⁻, by high-resolution LC-ESI-FTICR MS.

No. [a]	Name	Accurate m/z [d] [M–H]⁻	Molecular Formula [c] [M–H]⁻	Rt [b] (min)	Error [e] (ppm)
1	ibuprofen	205.12343	$C_{13}H_{17}O_2^-$	4.1	0.14
2	1-hydroxy IBP	221.11835	$C_{13}H_{17}O_3^-$	6.2	0.13
3	4-(1-hydroxy isobutyl)acetophenone	191.10765	$C_{12}H_{15}O_2^-$	4.6	−0.50

[a] Number used to identify each compound in the chromatograms of Figures 4 and 5. [b] Chromatographic retention time of compounds eluted under the experimental conditions described in Section 3. [c] Molecular formula of deprotonated compound. [d] Accurate m/z value of deprotonated molecules. [e] Mass error in parts per million = $10^6 \times$ (accurate mass − exact mass)/exact mass.

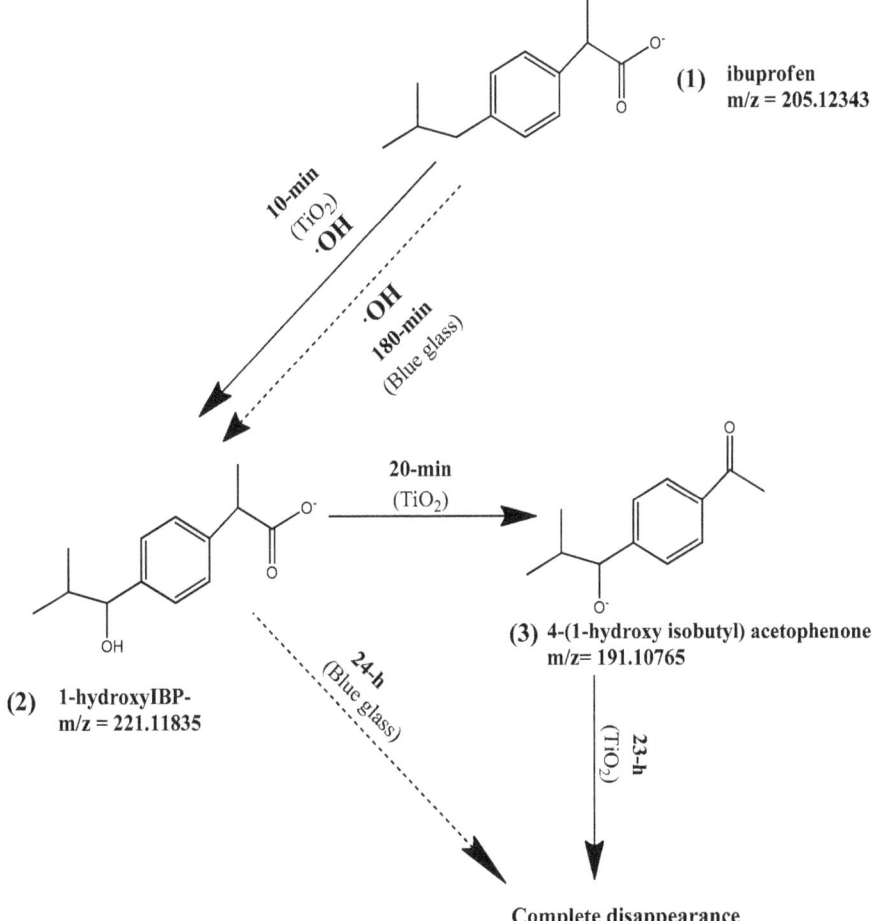

Figure 8. By-products generated by TiO₂ photocatalytic processes identified by LC-FTICR MS system in negative ion mode and proposed photodegradation pathway.

It is worth noting that, in the photodegradation method using TiO₂ immobilized on the active glass surface, only one by-product, compound 2, was detected.

3. Materials and Methods

3.1. Chemicals and Analytical Methods

Ibuprofen (MW, 206.3 g·mol^{-1}; pKa, 5.0) pure standard (purity, 99%) was purchased from Sigma Aldrich (Munich, Germany); acetonitrile, formic acid, and water for analysis were HPLC grade and purchased from Sigma Aldrich; TiO$_2$ Degussa P-25 was a kind compliment from Evonik Industries (Steinheim, Germany); TiO$_2$-coated active glass (Figure 9) was obtained as a gift from Pilkington (UK) (Sheel et al. 1998). PTFE (polytetrafluoroethylene) filters, 0.2 µm pore size, filter-Ø: 25mm, were purchased from Macherey-Nagel GmbH & Co. KG (Duren, Germany). Daily fresh working solutions were prepared using ultra-pure water from a bi-distilled purification system.

Figure 9. TiO$_2$-coated blue glass.

To avoid microbial contamination, all glass apparatus was heat-sterilized by autoclaving for 60 min at 121 °C before use. Aseptic handling materials and laboratory facilities were used throughout the study to maintain sterility.

IBP concentrations were monitored using high-performance liquid chromatography (HPLC) (1200 series, Agilent Technologies, Santa Clara, USA) equipped with an Eclipse XDB-C18 (3 µm particle size, 4.6 × 150 mm) column (Phenomenex, Torrance, USA) using a diode array detector (DAD) at a wavelength of 230 nm. The mobile phase consisted of 40% of 1% formic acid solution/60% acetonitrile. The flow rate was 1.0 mL·min^{-1}. Several aqueous solutions (from 0.5 to 25.0 mg·L^{-1}) of IBP were filtered, and 20 µL of the filtrate was injected and analyzed. Peak areas vs. concentration of IBP were plotted, and the calibration curve was obtained with a determination coefficient (R^2) of 0.9986. The limit of detection (LOD) of IBP for this method (using DAD) was 0.2 mg·L^{-1}, and the limit of quantitation (LOQ) was 0.6 mg·L^{-1}. The identification of IBP photoproducts was performed using the LC-FTICR MS system (Thermo Fisher Scientific, Bremen, Germany), in the same separation conditions. Negative ion ESI-MS mode was used for the detection of the compounds of interest. Full-scan experiments were performed in the ICR trapping cell in the range m/z 50–1000. Mass-to-charge ratio signals (m/z) were acquired as profile data at a resolution of 100,000 full width at half maximum (FWHM) at m/z 400. The limit of detection for mass spectrometric method was a few pmols.

The photodegradation experiments were performed using a solar simulator device Heraeus Sun-test CPS+ (Atlas, Chicago, USA), equipped with a 1500-W xenon arc lamp protected with a quartz filter (total passing wavelength: 280 nm < λ < 800 nm). The irradiation chamber was maintained at 20 °C by circulating water from a thermostatic bath and through a conditioned airflow.

3.2. Characterization of the TiO$_2$-Coated Active Glass

The active glass was obtained via a coating process using a nanocrystalline film of TiO$_2$ on a 4-mm-thin glass sheet. Some cross-sections obtained from theTiO$_2$-coated active glass were analyzed. The scanning electron analysis for the TiO$_2$-coated active glass was accomplished using a scanning electron microscope (SEM) of LEO model EVO50XVP, Carl Zeiss AG-EVO® 50 Series (Germany). The thin sections were grafted with 30-nm-thick carbon films. Semi-quantitative analyses of the elemental composition of the different layers were obtained using a Ge ED Oxford-Link detector equipped with a super atmosphere thin window. Operating conditions of the SEM were as follows: 15 kV accelerating potential, 500 pA probe current, and about 10 mm working distance (WD). Thin sections of glass were prepared by the Department of Health and Environmental Science, Bari University. Samples were embedded in resin epoxy plugs and then polished.

3.3. Photolysis Experiment

Aqueous IBP solution of initial concentration 25 mg·L^{-1} was prepared by dissolving a determined quantity of standard IBP in ultrapure water. The measured pH of the solution was 4.5. The photolysis treatment was carried out in a glass Pyrex®batch reactor closed at the top with a quartz cover. IBP solution (250 mL) was placed into the reactor; then, thereactor was placed into the irradiation oven inside the solar simulator, which reproduced the spectral distribution of natural solar irradiation. The IBP solution was continuously remixed during the experiment by magnetic stirring, and samples were taken (1 mL for each sample) at determined intervals and then analyzed using the HPLC system according to the analysis method in Section 3.1. Three experiments of direct photolysis were performed in triplicate.

3.4. Photocatalysis Experiment with TiO$_2$Powder

A solution of ibuprofen was prepared as described in Section 3.3, but 50 mg of TiO$_2$ powder (0.2 g·L^{-1}) was added as the optimized amount in the reactor vessel. The aqueous suspension was mixed continuously in the dark for 2 h to ensure that the adsorption equilibrium of IBP on the catalyst surface was reached; then, the reactor was transferred into the solar simulator and exposed to solar irradiation, and samples were taken (1 mL for each sample) at determined intervals, then filtered and analyzed by HPLC. Three experiments were performed in triplicate.

3.5. Photocatalysis Experiment with TiO$_2$Immobilized on Active Glass

Seven active glass sheets were placed vertically to cover the full surface of the inner wall of the reactor; then, the solution of IBP, prepared as in the previous two experiments, was added, and the reactor was transferred into the solar simulator and exposed to xenon lamp irradiation with continuous mixing; samples were taken (1 mL for each sample) at determined intervals, and then filtered and analyzed. Three experiments were performed in triplicate.

4. Conclusions

IBP is very stable under direct photolysis conditions due to its high chemical stability and low molar adsorption coefficient in the range of wavelengths provided by solar irradiation. On the other hand, our experiments showed that effective destruction of IBP and its photoproducts is possible by photocatalysis in the presence of TiO$_2$ powder suspension or using TiO$_2$immobilizedon the surface of active glass. Two intermediate photo-products were detected and identified by LC-FTICR MS. As there is substantial equivalence in the long-term efficacy of photocatalys is in the presence of TiO$_2$ both as powder suspension and as glass coating, the use of active glass instead of TiO$_2$ suspension could be a promising technique for the removal of pharmaceutical residues such as IBP and its photoproducts from aquatic environments not requiring the recovery of the catalyst after photodegradation. To increase the effectiveness of the technique described herein, a modification of TiO$_2$immobilization on the glass surface using supports with a more complex geometry is essential.

Author Contributions: S.K. This work is a part of his PhD thesis (Chapter 4). Therefore all experiments and results and manuscript writing were referring to his own efforts; J.H.S. Supervision and reviewing results, reviewing the manuscript English writing; F.L. LCMS analysis; L.S. SEM analysis; S.A.B. Supervision and reviewing results, reviewing the manuscript English writing; R.K. Reviewing the manuscript English writing. All authors have read and agreed to the published version of the manuscript.

Funding: This work was supported by the European Union in the framework of the Project "Diffusion of nanotechnology-based devices for water treatment and recycling; NANOWAT" (ENPI CBC MED I-B/2.1/049, Grant No. 7/1997).

Acknowledgments: Many thanks to Jawad H. Shoqueir, the head of Soil and hydrology Lab at Al-Quds University, for his support to partially cover the publication fee from his own budget. Results reported in this article were partially presented by Samer Khalaf at the Second International Conference on Recycle and Reuse, 4–6 June 2014, Istanbul, Turkey and published in the book of abstracts.

Conflicts of Interest: The authors declare no conflict of interest.

References

1. Jorgensen, S.E.; Halling-Sorensen, B. Drugs in the environment. *Chemosphere* **2000**, *40*, 691–699. [CrossRef]
2. Oron, G.; Gillerman, L.; Bick, A.; Buriaskovsky, N.; Gargir, M.; Dolan, Y.; Mnoar, L.; Katz, L.; Hagin, J. Membrane technology for advanced wastewater reclamation for sustainable agriculture production. *Desalination* **2008**, *218*, 170–180. [CrossRef]
3. Renge, V.C.; Khedkar, S.V.; Thanvi, N.J. Photocatalytic-oxidation and reactors. *Int. J. Adv. Eng. Technol.* **2012**, *3*, 31–35.
4. Ahmed, S.; Rasul, M.G.; Martens, W.N.; Brown, R.; Hashib, M.A. Heterogeneous photocatalytic degradation of phenols in wastewater: A review on current status and developments. *Desalination* **2010**, *261*, 3–18. [CrossRef]
5. Madhavan, J.; Grieser, F.; Ashokkumar, M. Combined advanced oxidation processes for the synergistic degradation of ibuprofen in aqueous environments. *J. Hazard. Mater.* **2010**, *178*, 202–208. [CrossRef] [PubMed]
6. Richardson, M.L.; Bowron, J.M. The fate of pharmaceutical chemicals in the aquatic environment. *J. Pharm. Pharmacol.* **1985**, *37*, 1–12. [CrossRef] [PubMed]
7. Santos, J.L.; Aparicio, I.; Alonso, E. Occurrence and risk assessment of pharmaceutically active compounds in wastewater treatment plants: A case study: Seville city (Spain). *Environ. Int.* **2007**, *33*, 596–601. [CrossRef]
8. Klavarioti, M.; Mantzavinos, D.; Kassinos, D. Removal of residual pharmaceuticals from aqueous systems by advanced oxidation processes. *Environ. Int.* **2009**, *35*, 402–417. [CrossRef]
9. Karaman, R.; Khamis, M.; Qurie, M.; Halabieh, R.; Makharzeh, I.; Manassra, A.; Abbadi, J.; Qtait, A.; Bufo, S.A.; Nasser, A.; et al. Removal of diclofenac potassium from wastewater using clay-micelle complex. *J. Environ. Technol.* **2012**, *33*, 1279–1287. [CrossRef]
10. Khalaf, S.; Rimawi, F.; Khamis, M.; Zimmerman, D.; Shuali, U.; Nir, S.; Scrano, L.; Bufo, S.A.; Karaman, R. Efficiency of advanced wastewater treatment plant system and laboratory-scale micelle-clay filtration for the removal of ibuprofen residues. *J. Environ. Sci. Health Part B* **2013**, *48*, 814–821. [CrossRef]
11. Khalaf, S.; Rimawi, F.; Khamis, M.; Nir, S.; Bufo, S.A.; Scrano, L.; Mecca, G.; Karaman, R. Efficiency of membrane technology, activated charcoal, and a micelle-clay complex for removal of the acidic pharmaceutical mefenamic acid. *J. Environ. Sci. Health Part A* **2013**, *48*, 1655–1662. [CrossRef] [PubMed]
12. Qurie, M.; Khamis, M.; Malek, F.; Nir, S.; Bufo, S.A.; Scrano, L.; Karaman, R. Stability and removal of naproxen and its metabolite by advanced membrane wastewater treatment plant and micelle-clay complex. *Clean Soil Air Water* **2013**, *42*, 594–600. [CrossRef]
13. Sulaiman, S.; Khamis, M.; Nir, S.; Lelario, F.; Scrano, L.; Bufo, S.A.; Karaman, R. Stability and removal of dexamethasone sodium phosphate from wastewater using modified clays. *J. Environ. Technol.* **2014**, *35*, 1945–1955. [CrossRef] [PubMed]
14. Andreozzi, R.; Raffaele, M.; Nicklas, P. Pharmaceuticals in STP effluents and their solar photodegradation in aquatic environment. *Chemosphere* **2013**, *50*, 1319–1330. [CrossRef]
15. Inoue, M.; Masuda, Y.; Okada, F.; Sakurai, A.; Takahashi, I.; Sakakibara, M. Degradation of bisphenol A using sonochemical reactions. *Water Res.* **2008**, *42*, 1379–1386. [CrossRef]
16. Stathis, I.; Hela, D.G.; Scrano, L.; Lelario, F.; Emanuele, L.; Bufo, S.A. Novel imazethapyr detoxification applying advanced oxidation processes. *J. Environ. Sci. Health Part B* **2011**, *46*, 449–453.
17. Okamoto, K.I.; Yamamoto, Y.; Tanaka, H.; Tanaka, M. Heterogeneous photocatalytic decomposition of phenol over TiO_2 powder. *Bull. Chem. Soc. Jpn.* **1985**, *58*, 2015–2022. [CrossRef]
18. Phanikrishna Sharma, M.V.; DurgaKumari, V.; Subrahmanyam, M. TiO_2 supported over SBA-15: An efficient photocatalyst for the pesticide degradation using solar light. *Chemosphere* **2008**, *73*, 1562–1569. [CrossRef]

19. Zhu, X.; Yuan, C.; Bao, Y.; Yang, Y.; Wu, Y. Photocatalytic degradation of pesticide pyridaben on TiO$_2$ particles. *Int. Nano Lett.* **2005**, *229*, 95–105. [CrossRef]
20. Serpone, N.; Pellizetti, E. *Photocatalysis: Fundamentals &Applications*; John Wiley & Sons, Inc.: New York, NY, USA, 1989; pp. 604–634.
21. Stylidi, M.; Kondarides, D.I.; Verykios, X.E. Pathways of solar light-induced photocatalytic degradation of azo dyes in aqueous TiO$_2$ suspensions. *Appl. Catal. B Environ.* **2003**, *40*, 271–286. [CrossRef]
22. Mills, A.; Lepre, A.; Elliott, N.; Bhopal, S.; Parkin, I.P.; O'Neill, S.A. Characterisation of the photocatalyst Pilkington Activ(TM): A reference film photocatalyst? *J. Photochem. Photobiol. A* **2003**, *160*, 213–224. [CrossRef]
23. Parkin, I.P.; Palgrave, R.G. Self-cleaning coatings. *J. Mater. Chem.* **2005**, *15*, 1689–1698. [CrossRef]
24. Sheel, D.W.; McCurdy, R.J.; Hurst, S.J. Method of Depositing Tinoxideand Titanium Oxide Coatings on Flat Glass and the Resultingcoated Glass. Patent Application WO 98/06675, 19 February 1998.
25. Murov, S.L.; Carmichael, I.; Hug, G.L. *Handbook of Photochemistry*; Marcel Dekker: New York, NY, USA, 1993; pp. 82–99.
26. Volman, D.H.; Seed, J.R. The photochemistry of uranyl oxalate. *J. Am. Chem. Soc.* **1964**, *86*, 5095–5098. [CrossRef]
27. Scrano, L.; Bufo, S.A.; Perucci, P.; Meallier, P.; Mansour, M. Photolysis and hydrolysis of rimsulfuron. *Pestic. Sci.* **1999**, *55*, 955–961. [CrossRef]
28. Snedecor, G.W.; Cochran, W.G. *Statistical Methods*, 8th ed.; Iowa State University Press: Ames, IA, USA, 1989.
29. Scrano, L.; Bufo, S.A.; Emmelin, C.; Meallier, P. Abiotic Degradation of the Herbicide Rimsulfuron on Minerals and Soil. In *Environmental Chemistry: Green Chemistry and Pollutants in Ecosystems*; Lichtfouse, E., Schwarzbauer, J., Robert, D., Eds.; Springer: Berlin/Heidelberg, Germany, 2005; pp. 505–515.
30. Braz, F.S.; Silva, M.R.A.; Silva, F.S.; Andrade, S.J.; Fonseca, A.L.; Kondo, M.M. Photocatalytic Degradation of ibuprofen using TiO$_2$ and ecotoxicological assessment of degradation intermediates against daphnia similis. *J. Environ. Protect.* **2014**, *5*, 620–626. [CrossRef]
31. Chu, W.; Choy, W.K.; So, T.Y. The effect of solution pH and peroxide in the TiO2-induced photocatalysis of chlorinated aniline. *J. Hazard. Mater.* **2007**, *141*, 86–91. [CrossRef]
32. Chong, M.N.; Jin, B.; Chow, C.W.K.; Saint, C. Recent Developments in Photocatalytic Water Treatment Technology: A Review. *Water Resour.* **2010**, *44*, 2997–3027. [CrossRef]
33. Kosmulski, M. The significance of the difference in the point of zero charge between rutile and anatase. *Adv. Colloid Interface Sci.* **2002**, *99*, 255–264. [CrossRef]
34. Ning, B.; Graham, N.; Zhang, Y.P.; Nakonechny, M.; El-Din, M.G. The degradation of endocrine disrupting chemicals by ozone and AOPs—A review. *J. Ozone Sci. Eng.* **2007**, *29*, 153–176. [CrossRef]

© 2020 by the authors. Licensee MDPI, Basel, Switzerland. This article is an open access article distributed under the terms and conditions of the Creative Commons Attribution (CC BY) license (http://creativecommons.org/licenses/by/4.0/).

Article

Photocatalytic Degradation of Chlorpyrifos with Mn-WO$_3$/SnS$_2$ Heterostructure

Charlie M. Kgoetlana, Soraya P. Malinga and Langelihle N. Dlamini *

Department of Chemical Sciences, University of Johannesburg, Doornfontein Campus, P.O. Box 17011, Doornfontein, Johannesburg 2028, South Africa; kgoetlanacm@gmail.com (C.M.K.); smalinga@uj.ac.za (S.P.M.)
* Correspondence: lndlamini@uj.ac.za; Tel.: +27-011-559-6945

Received: 21 May 2020; Accepted: 4 June 2020; Published: 21 June 2020

Abstract: Tungsten trioxide (WO$_3$) is a photocatalyst that has gained interest amongst researchers because of its non-toxicity, narrow band gap and superior charge transport. Due to its fast charge recombination, modification is vital to counteract this limitation. In this paper, we report on the fabrication of Mn-doped WO$_3$/SnS$_2$ nanoparticles, which were synthesised with the aim of minimising the recombination rates of the photogenerated species. The nanomaterials were characterised using spectroscopic techniques (UV-Vis-diffuse reflectance spectroscopy (DRS), Raman, XRD, photoluminescence (PL) and electrochemical impedance spectroscopy (EIS)) together with microscopic techniques (FESEM-EDS and high resolution transmission electron microscopy selected area electron diffraction (HRTEM-SAED)) to confirm the successful formation of Mn-WO$_3$/SnS$_2$ nanoparticles. The Mn-doped WO$_3$/SnS$_2$ composite was a mixture of monoclinic and hexagonal phases, confirmed by XRD and Raman analysis. The Mn-WO$_3$/SnS$_2$ heterojunction showed enhanced optical properties compared to those of the un-doped WO$_3$/SnS$_2$ nanoparticles, which confirms the successful charge separation. The Brunauer–Emmett–Teller (BET) analysis indicated that the nanoparticles were mesoporous as they exhibited a Type IV isotherm. These nanomaterials appeared as a mixture of rectangular rods and sheet-like shapes with an increased surface area (77.14 m^2/g) and pore volume (0.0641 cm^3/g). The electrochemical measurements indicated a high current density (0.030 mA/cm^2) and low charge transfer resistance (157.16 Ω) of the Mn-WO$_3$/SnS$_2$ heterojunction, which infers a high charge separation, also complemented by photoluminescence with low emission peak intensity. The Mott–Schottky (M-S) plot indicated a positive slope characteristic of an n–n heterojunction semiconductor, indicating that electrons are the major charge carriers. Thus, the efficiency of Mn-WO$_3$/SnS$_2$ heterojunction photocatalyst was monitored for the degradation of chlorpyrifos. The effects of pH (3–9), catalyst loading (0.1–2 g) and initial chlorpyrifos concentration (100 ppb–20 ppm) were studied. It was observed that the degradation was purely due to photocatalysis, as no loss of chlorpyrifos was observed within 30 min in the dark. Chlorpyrifos removal using Mn-WO$_3$/SnS$_2$ was performed at the optimum conditions of pH = 7, catalyst loading = 1 g and chlorpyrifos concentration = 1000 ppb in 90 min. The complete degradation of chlorpyrifos and its major degradation by-product 3,5,6-trichloropyridin-2-ol (TCP) was achieved. Kinetic studies deduced a second order reaction at 209 × 10^{-3} M^{-1}s^{-1}.

Keywords: heterojunction; charge separation; photocatalysis; chlorpyrifos

---

1. Introduction

The fabrication and modification of photocatalysts has sparked interest amongst researchers due to their wide applications. Photocatalysts are used in applications ranging from water splitting, the degradation of pollutants in water, gas sensing and optoelectronic devices [1]. These can be n-type (electrons are the major charge carriers) or p-type (holes are the major charge carriers) photocatalysts [2]. The most widely studied photocatalysts are TiO_2, WO_3 (*n*-type) and ZnO, CdS (*p*-type) [3,4]. The photocatalytic efficiency of these materials is limited to a certain extent, primarily due to two major limitations. Firstly, they are prone to fast electron–hole recombination, which reduces the photocatalytic reactivity of the semiconductor. Secondly, they have wide band gaps that absorb only in the ultraviolet (UV) region, which accounts for 4% of the solar spectrum [5].

Modifications of photocatalysts to suit specific applications have been proposed. These include the use of metal dopants to form Schottky barriers and fusion with other semiconductor photocatalysts, resulting in heterojunctions [2]. The metal dopants that have been employed include magnesium (Mg), manganese (Mn), copper (Cu) and yttrium (Y) [6–9]. The metal doping of photocatalysts results in shifting the absorption band edge of the material to absorb the readily available visible region of the solar spectrum. They also separate photogenerated charges by forming electron traps, although a high concentration of the metal dopant may result in the creation of recombination centres, which leads to an increased recombination rate.

Heterojunctions that have shown enhanced optical and photocatalytic properties include $BiVO_4/WO_3$, CdS/ZnO and TiO_2/SnO_2 [10–13]. The formation of Type II heterojunctions using two different photocatalysts is sufficient to reduce the recombination rate of photogenerated charges. This occurs by the accumulation of photoexcited electrons in the conduction band (CB) of one semiconductor while photogenerated holes accumulate in the valence band (VB) of another semiconductor in the heterojunction system, which effectively leads to charge separation. Therefore, photo-oxidation and photo-reduction occur in different semiconductor surfaces of the heterojunction system due to the different migration points of the charges [14].

Tungsten trioxide (WO_3) is a visible light photocatalyst with a band gap energy of 2.5–2.8 eV [15]. It is classified as an *n*-type semiconductor, wherein electrons are the major carriers. Due to its narrow band gap, WO_3 absorbs light radiation in the visible range, and it has been used in a wide range of applications such as fuel production and combating water pollution [16,17]. This semiconductor exists in different polymorphs, which include monoclinic, triclinic, tetragonal and orthorhombic. The monoclinic phase of WO_3 is the most stable and most photocatalytic compared to all the other phases.

Like most photocatalysts, WO_3 suffers from limitations such as high electron–hole charge recombination. To overcome the intrinsic limitation of pristine WO_3, different methods have been used, including metal doping and loading another semiconductor photocatalyst to form a heterojunction [10,18,19].

We, however, report the synthesis and characterisation of a material that fuses metal doping and a heterojunction (Mn-WO_3/SnS_2) that exhibits improved optical properties and minimises electron–hole recombination compared to current photocatalysts. Owing to the charge mobility in SnS_2 to efficiently facilitate electron transfer to the WO_3 CB, there results a high number of electrons for the oxidation reaction and separated holes that accumulate on the VB of SnS_2. The Mn^{2+} ions also separate charges by trapping electrons in the WO_3, thereby increasing their lifetime, and act as reaction centres. This photocatalyst can be applied in water remediation, energy production and sensing. Thus, this study assessed the photo-efficiency of the heterostructure in the photodegradation of chlorpyrifos, an organophosphate pesticide.

Organophosphate pesticides have been used extensively in South Africa and the world at large due to their ability to combat a vast spectrum of pests [20,21]. Chlorpyrifos (O,O-diethyl O-[3,5,6, -trichloro-2-pyridyl] phosphorothionate) (CPF) is an organophosphorus pesticide extensively used in agricultural and domestic applications [22]. Chlorpyrifos agricultural application occurs throughout the year for a variety of fruits and vegetables [23]. It, however, does not readily dissolve in water, yet adsorbs strongly to soil particles.

Chlorpyrifos is an enzyme acetylcholinesterase inhibitor and persistent pesticide pollutant. It is a class II (moderately hazardous pesticide) pollutant, with a half-life of 60 days [24]. The pesticide is toxic to humans and other animals when ingested or inhaled; this is attributed to its lipophilic nature.

It causes delayed peripheral neuropathy in humans and badly affect neuro-development in children at high doses [25,26].

Due to the numerous human and environmental effects caused by chlorpyrifos, different ways to remove this pesticide from the environment have been studied. These include advanced oxidation processes and biological treatment (with fungal and bacterial strains).

Bacterial strains have displayed high chlorpyrifos removal from water of up to 98% [27]. Though it is efficient, the method is strenuous, as bacteria require controlled specific conditions such as pH and temperature and a host for optimal function. On the other hand, Ismail et al. [24] in 2013 reported the use of advanced oxidation processes (AOPs) that yield 100% removal of chlorpyrifos by using ^{60}Co γ-rays of 30–575 Gy [24].

However, γ-rays are harmful to human health; therefore, this led to the implementation of a better and safer method requiring the use of a photocatalyst to degrade chlorpyrifos under light irradiation. To date, zinc and titanium oxides have been used to degrade chlorpyrifos. The results were satisfactory, with up to 95% chlorpyrifos removal for TiO_2 and 85% for ZnO under UV light [28]. However, the photocatalysts suffer from charge recombination and the use of UV light is not viable due to the insufficient amount of UV available (4%). Therefore, visible light-absorbing photocatalysts were discovered such as $BiVO_2$, SnS_2 and WO_3.

To the best of our knowledge, no work has been reported to date on the fabrication of a metal-doped heterojunction (Mn-WO_3/SnS_2) photocatalyst.

2. Results and Discussion

2.1. X-ray Diffraction and Raman Analyses

The phase and crystallographic properties of the nanomaterials were elucidated using XRD and Raman spectroscopy. Figure 1a shows the XRD pattern of WO_3, which confirms the monoclinic nature of the WO_3 (m-WO_3). The m-WO_3 was indexed and matched to the miller indices (002), (020), (200), (120), (112), (022), (202), (122), (222), (004), (040), (400), (042) and (420) (JCPDS Card No. 00-043-1035). Doping the m-WO_3 with Mn^{2+} did not distort the phase of the WO_3, which implies that it had been intrinsically inserted into the WO_3 crystal lattice as depicted in Figure 1b. The XRD pattern of the WO_3/SnS_2 heterojunction showed the presence of both monoclinic (m-WO_3) and hexagonal (h-WO_3) phases, and the hexagonal phase of SnS_2 could be indexed (JCPDS Card No. 00-023-0677), as shown in Figure 1c. Again, the structural integrity of the manganese-doped WO_3/SnS_2 heterojunction (Figure 1d) was not distorted by the incorporation of Mn^{2+} in the system. The average crystallite sizes of the nanomaterials were determined using the Debye–Scherrer equation and are tabulated in Table S1, all with an average size of 40 nm, with SnS_2 having a crystallite size of less than 20 nm.

The nature of the phases was further confirmed with Raman analysis. Figure 2a illustrates Raman bands at 717 and 818 cm^{-1} and less intense bands at 212 and 313 cm^{-1} corresponding to O-W-O stretching and bending in the molecule, respectively, which confirms a monoclinic WO_3; this finding was also reported by Simelane et al. 2017 and Xie et al. 2012 [3,15]. As in XRD analyses, the doping of Mn^{2+} had no effect on the phase of WO_3, as depicted in Figure 2b. No secondary bands resulting from Mn-oxides were observed. The heterojunction (WO_3/SnS_2) was successfully formed and confirmed by the Raman band at 317 cm^{-1} corresponding to the A_{1g} mode of hexagonal phase SnS_2 as observed by Ma et al. 2015 (Figure 2c) [29]. Figure 2d displays the Raman band of Mn-WO_3/SnS_2 with no distortion due to Mn^{2+} and SnS_2. Therefore, the Raman band in Figure 2e corresponds to the pristine hexagonal phase of SnS_2 due to the A_{1g} band at 317 cm^{-1}.

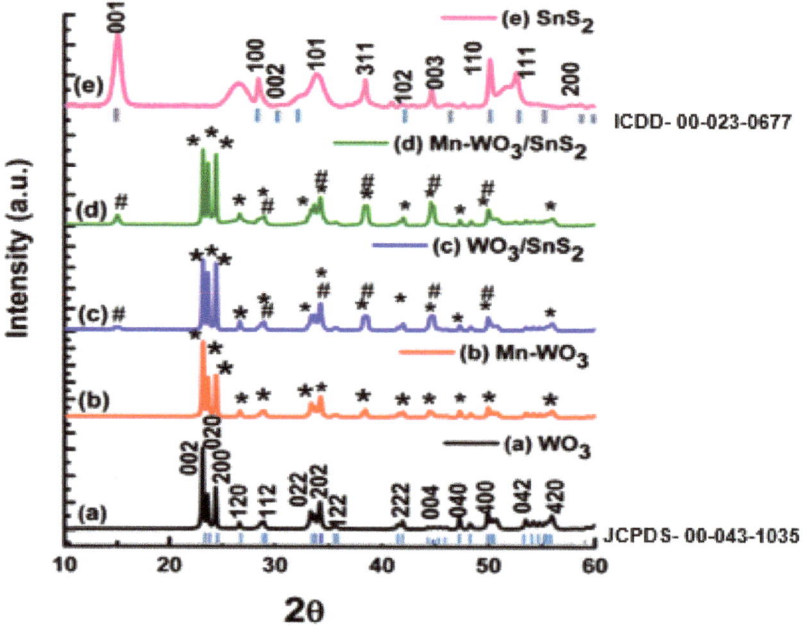

Figure 1. XRD patterns of (**a**) WO_3, (**b**) $Mn-WO_3$, (**c**) WO_3/SnS_2, (**d**) $Mn-WO_3/SnS_2$ and (**e**) SnS_2.

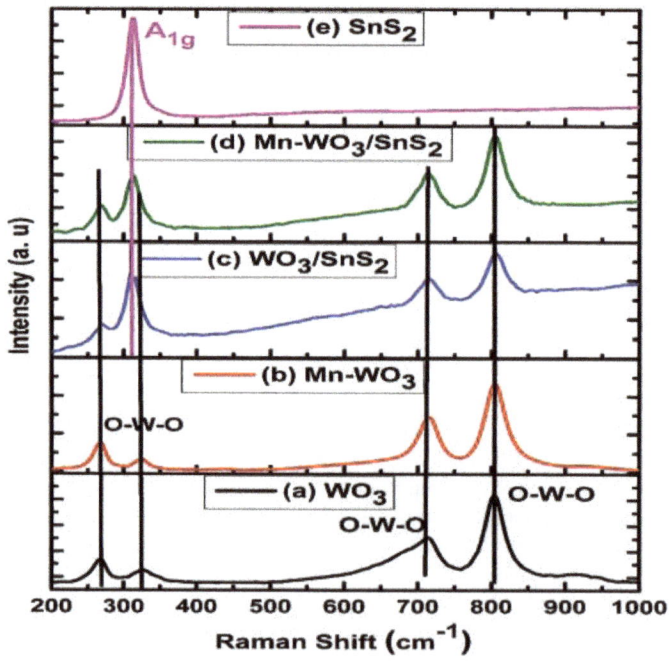

Figure 2. Raman spectra of (**a**) WO_3, (**b**) $Mn-WO_3$, (**c**) WO_3/SnS_2, (**d**) $Mn-WO_3/SnS_2$ and (**e**) SnS_2.

2.2. Morphological Studies

The morphological studies were conducted using microscopic techniques such as FESEM and HRTEM. Figure 3a is the FESEM image of pristine m-WO$_3$ with rectangular sheets, rods and cubes, and the composition was confirmed by EDX (inset). The shapes of the nanomaterials did not change upon the insertion of Mn^{2+} or the formation of the heterojunction as illustrated in Figure 3b,c and Figures S2 and S3. The EDX spectra displayed the elemental composition of the respective heterojunctions (WO$_3$/SnS$_2$) and Mn-WO$_3$/SnS$_2$ (inset).

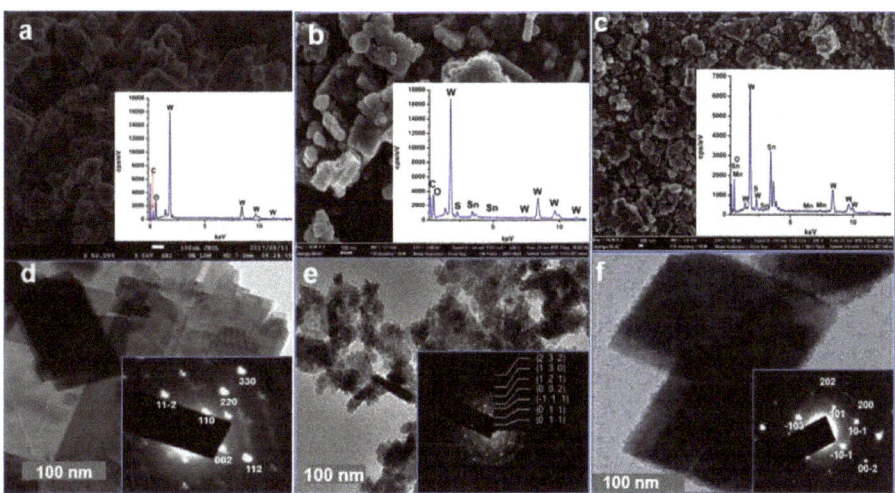

Figure 3. FESEM images (inset is the corresponding energy-dispersive X-ray (EDX) spectrum) of (**a**) pristine WO$_3$, (**b**) WO$_3$/SnS$_2$ and (**c**) Mn-WO$_3$/SnS$_2$, and TEM images (inset is the corresponding selected area electron diffraction (SAED) image) of (**d**) WO$_3$, (**e**) WO$_3$/SnS$_2$ and (**f**) Mn-WO$_3$/SnS$_2$.

The HRTEM image of m-WO$_3$ also showed rectangular sheets and rods (Figure 3d) and was further elucidated using SAED (inset) obtained through a 1–10 zone axis. The spots were indexed to (002), (220) and (112) corresponding to monoclinic WO$_3$ as confirmed by XRD analysis. The HRTEM images (Figure 3e,f) displayed rectangular rods and sheet-like shapes as observed in Figure 3d, which implies that no shape distortion had occurred through metal doping and the formation of the heterojunction. The SAED image (inset) displays spots and rings characteristic of monoclinic WO$_3$ and the SnS$_2$ hexagonal phase, respectively. Furthermore, the SAED image (inset) illustrates spot (202, 200) and ring (101, −103) indices corresponding to the WO$_3$ monoclinic phase and SnS$_2$ hexagonal phase, respectively, captured through a 0–10 zone axis using the CrysTBox software [30] (Figure 3f). All the SAED indices correspond to the reported XRD patterns, which further confirms the successful formation of our nanomaterial. In the Mn-WO$_3$/SnS$_2$, the estimated percentage of Mn was 2.5%, with 47.5% of WO$_3$ and 50% of SnS$_2$.

2.3. Optical Properties

Ultraviolet-visible spectroscopy in diffuse reflectance mode was used to determine the optical properties of the synthesised nanoparticles. All the synthesised nanomaterials showed a shift of absorption to be in the visible region, which is in abundance. Pristine m-WO$_3$ displayed a band gap of 2.71 eV, with a corresponding absorption wavelength of 466 nm, as shown in Figure 4a and Figure S4, respectively. The value obtained agrees with the value reported by Simelane et al. in 2017 [15]. The insertion of Mn^{2+} in the m-WO$_3$ lattice introduced impurities and thus a shift in the Fermi level below the conduction band, promoting the red-shifting of WO$_3$ on the absorption spectrum (Figure 4b);

this was observed by Harshulkhan et al. in 2017 using magnesium as a dopant [6]. The band gap of the WO$_3$ was reduced upon the insertion of Mn and decreased further after the formation of a heterojunction with SnS$_2$ (Figure 4a,b,d,e). The Mn-doped heterojunction had the lowest band gaps (2.08 eV and 2.34 eV) amongst the nanomaterials (the others were WO$_3$, SnS$_2$, Mn-WO$_3$ and WO$_3$/SnS$_2$), which correspond to a high light absorption wavelength (red-shift) (Figure S4); this was due to visible light absorption enhancement by both the Mn ion and SnS$_2$.

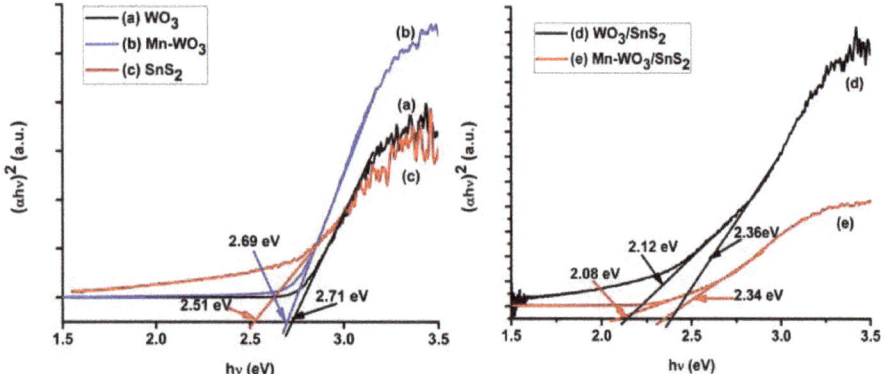

Figure 4. Tauc plots indicating the band gaps of (**a**) WO$_3$, (**b**) Mn-WO$_3$, (**c**) SnS$_2$, (**d**) WO$_3$/SnS$_2$ and (**e**) Mn-WO$_3$/SnS$_2$.

The diagram in Figure 5 illustrates the change in the band edge potential of the synthesised semiconductor photocatalysts. The valence band edge potential (E_{VB}) and the conduction band edge potential (E_{CB}) were calculated using Equations (5) and (6). A slight decrease in both E_{CB} and E_{VB} was observed during the introduction of Mn and fusion with SnS$_2$ (Mn-WO$_3$/SnS$_2$). The conduction band edge potential shifted to be more positive (by 0.2 eV), and the valence band edge potential moved to a less positive potential (by 0.2 eV); this was due to the insertion of an ion with a high ionic radius, which reduces the band gap by pulling the band edges closer, resulting in band edge shifts. The change in the position of the band edges enhances the absorption wavelength of the material. The heterojunction (Mn-WO$_3$/SnS$_2$) enhances charge separation by the movement of electrons from the SnS$_2$ CB to the WO$_3$ CB through the interface, thereby leaving holes in the VB of the SnS$_2$. This effectively separates the electrons and holes as they accumulate in the CB of WO$_3$ and the VB of SnS$_2$, respectively.

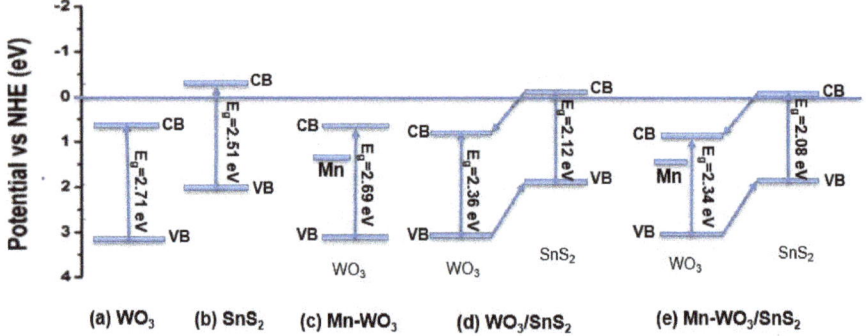

Figure 5. Diagram of the band *gap, valence band* and *conduction* band *edge positions* vs. the NHE of the (**a**) WO$_3$, (**b**) SnS$_2$, (**c**) Mn-WO$_3$, (**d**) WO$_3$/SnS$_2$ and (**e**) Mn-WO$_3$/SnS$_2$ photocatalysts.

2.4. Electrochemical and Photoluminescence Measurements

The electrochemical impedance spectroscopy (EIS) measurements were carried out to study the interfacial reactions occurring between the photoelectrode and the electrolyte. Figure 6a illustrates the EIS spectrum (Nyquist plot) with a suppressed semicircle with a large diameter. At low frequency, the current density is in phase with the potential deviation of the system, resulting in a straight line at an angle of 45° to the X-axis. The large diameter of the semicircle at high frequency corresponds to the high charge transfer impedance of WO_3. This relates to the high charge recombination rate as observed in Figure 6a. The charge transfer impedance was reduced after WO_3 was doped with the Mn^{2+} ion (Figure 6b), due to the reduced charge recombination rate and increased charge mobility. This was due to the Mn^{2+} ions acting as charge collection sites, thereby serving as an electrical conduction pathway, allowing ion/electron mobility on the electrode [7]. The small diameter of the semicircle of the $Mn-WO_3/SnS_2$ spectrum indicates decreased electrode–electrolyte charge-transfer resistance/impedance compared to that in the WO_3/SnS_2, $Mn-WO_3$, SnS_2 and WO_3 in the 0.1 M Na_2SO_4 electrolyte. The sloping straight line in the low-frequency region corresponds to oxygen diffusion within the electrode (Figure 6a). The low charge transfer resistance of $Mn-WO_3/SnS_2$ arises from the enhanced charge carrier separation induced by the Mn^{2+} ion dopant and SnS_2 semiconductor heterojunction with WO_3. The charge transfer resistance and recombination rate decreased for WO_3, $Mn-WO_3$, SnS_2 and WO_3/SnS_2, with the lowest rate observed in the $Mn-WO_3/SnS_2$ (Figure 6a,b).

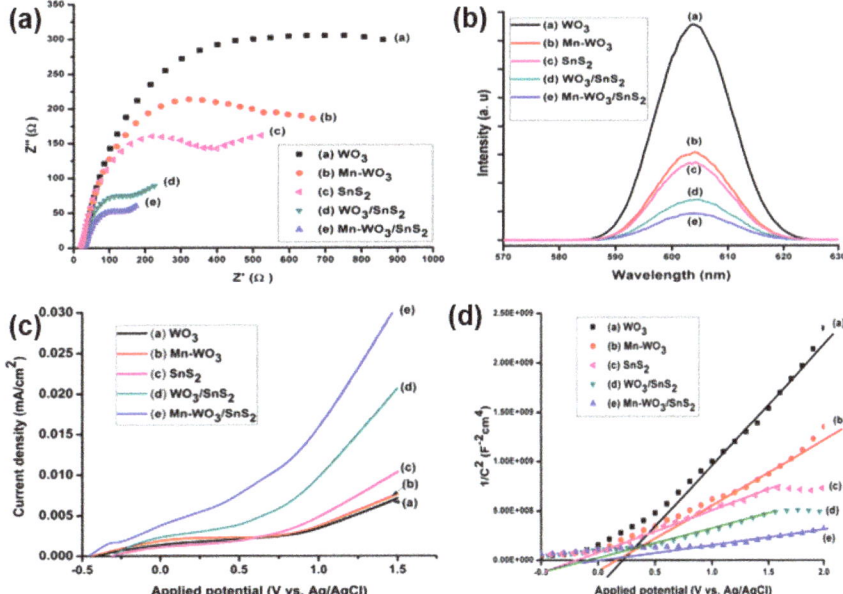

Figure 6. (a) Electrochemical impedance spectra (Nyquist plot), (b) photoluminescence spectra, (c) linear sweep voltammetry, and (d) Mott–Schottky plot of (a) WO_3, (b) $Mn-WO_3$, (c) SnS_2, (d) WO_3/SnS_2 and (e) $Mn-WO_3/SnS_2$.

Photoluminescence (PL) measurements (Figure 6b) support the EIS findings that the WO_3 has a strong PL intensity, which indicates high charge carrier recombination, and it was reduced by the introduction of Mn and fusion with SnS_2. A decrease in PL intensity was observed in the $Mn-WO_3/SnS_2$, indicating low charge carrier recombination, which implied that it would be a good photocatalyst in photocatalytic applications. This is attributed to the longer charge carrier lifetimes and enhanced charge carrier mobility provided by the Sn-S bond, thereby minimising the electron–hole recombination.

Upon the introduction of Mn^{2+} and SnS_2, the photocurrent density of WO_3 was observed to be improved by up to 0.030 mA/cm^2 for Mn-WO_3/SnS_2 NPs (Figure 6c). This implied that there was high electron flow between the photocatalyst and the electrolyte produced from the photocatalyst upon light irradiation. The Mn^{2+} acts as an electron sink and reaction side, which in turn supplies electrons for interfacial reactions, and upon illumination, SnS_2 helps in the production of electrons and their separation from holes, which increases the current density.

Mott–Schottky plots were used to study the interfacial capacitance of the nanomaterials. The positive slopes obtained from Figure 6d confirmed that the synthesised nanomaterials are all *n*-type semiconductors, which use electrons as major charge carriers. The positive slope for the heterojunction WO_3/SnS_2 and Mn-WO_3/SnS_2 NPs further inferred the formation of an *n–n* type heterojunction system. Upon the introduction of the Mn^{2+} and formation of the heterojunction, there was no significant change in the slope of the curves.

The flat-band potential (V_{fb}) was obtained by extrapolating a line on the slope of the graph to the *x*-intercept ($1/C^2=0$). The flat-band potentials were found to be 0.214 V, 0.159 V, −0.209 V, −0.103 V and −0.039 V, corresponding to WO_3, Mn-WO_3, SnS_2, WO_3/SnS_2 and Mn-WO_3/SnS_2, respectively. The flat-band potential in *n*-type semiconductors corresponds to the bottom of the conduction band of the semiconductor photocatalyst, which was observed to decrease upon doping and the formation of the heterojunction (Figure 6d). The obtained flat-band potential (V_{fb}) values were found to correspond to the calculated conduction band edge potentials (E_{CB}) from UV-Vis DRS.

The Randles equivalent circuit model was used to fit the obtained EIS data. The Randles equivalent circuit models (Figure 7a,b) corresponding to the graphs show that the impedance was a contribution of three forms of resistance, namely the solution resistance, the electrode resistance due to the film composition of the nanomaterials, and charge-transfer resistance occurring at the electrolyte–electrode interface.

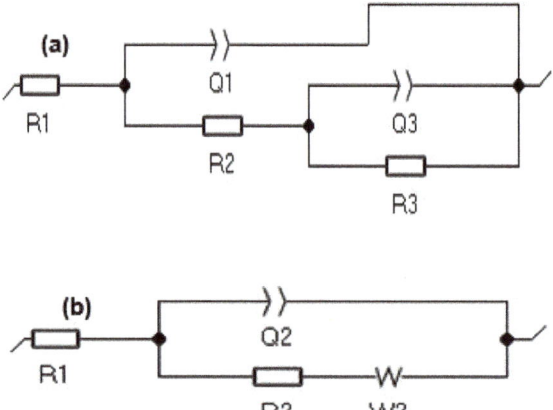

Figure 7. Randles equivalent circuit models corresponding to (a) WO_3 and Mn-WO_3, and (b) SnS_2, WO_3/SnS_2 and Mn-WO_3/SnS_2.

The Warburg impedance is due to solid-state ion diffusion during the electrochemical reaction. The Warburg element manifests itself in EIS spectra as a straight line with a slope of 45° in the low-frequency region. Different slopes of the straight-line part in the low-frequency region indicate that the electrodes have different Warburg impedances and solid-state ion diffusion behaviors.

The equivalent circuit model was obtained after fitting the data to the Randles model: R1 is the solution resistance, R2 is the thin layer resistance, R3 is the charge transfer resistance, W2 is the Warburg resistance and Q1, Q2 and Q3 are the constant phase elements. The Warburg impedance relates to solid-state ion diffusion during the electrochemical reaction in the solution. This favours photocatalytic activity by utilising the separated charges during the reaction and, consequently, reduces charge recombination. The slope of the Warburg transition line also indicates the reactivity of the nanoparticles. Furthermore, the charge transfer impedance was found to be 631.80, 498.50, 310.55, 173.65 and 157.16 Ω for WO_3, Mn-WO_3, SnS_2, WO_3/SnS_2 and Mn-WO_3/SnS_2, respectively.

2.5. BET Analysis

The BET analysis revealed that the nitrogen adsorption isotherms obtained for the nanoparticles were Type IV isotherms, according to the IUPAC (International Union of Pure and Applied Chemistry) classification (indicated in Figure S6). A Type IV isotherm is typical of mesoporous materials (IUPAC definition: pore size 2–50 nm), suggesting that the nanomaterials consist of agglomerates. The Mn-WO_3/SnS_2 nanoparticles were found to have the highest BET surface area (77.14 m^2/g) and pore volume, of 0.0641 cm^3/g, compared to pristine materials (Table 1). This suggests that Mn-WO_3/SnS_2 would have improved photocatalytic activity due to the adsorption capacity provided by its large specific surface area during photocatalysis. The large pore volume would allow the efficient trapping of pollutants during adsorption for degradation to take place.

Table 1. The specific surface area and pore volume of NPs.

Material	S_{BET} (m^2/g)	Pore Volume (cm^3/g)
WO_3	6.01	0.0276
Mn-WO_3	4.41	0.0294
WO_3/SnS_2	44.36	0.0514
Mn-WO_3/SnS_2	77.14	0.0641
Pristine SnS_2	99.72	0.0748

2.6. Surface Charge of Nanoparticles

The stability of the nanomaterials in suspensions was studied using the electrophoretic light scattering technique. The zeta potentials of the nanomaterials are illustrated in Figure 8. showing a steady but gradual change in zeta potential from positive to negative as the pH increased from 2 to 11 for all the photocatalysts (Figure 8).

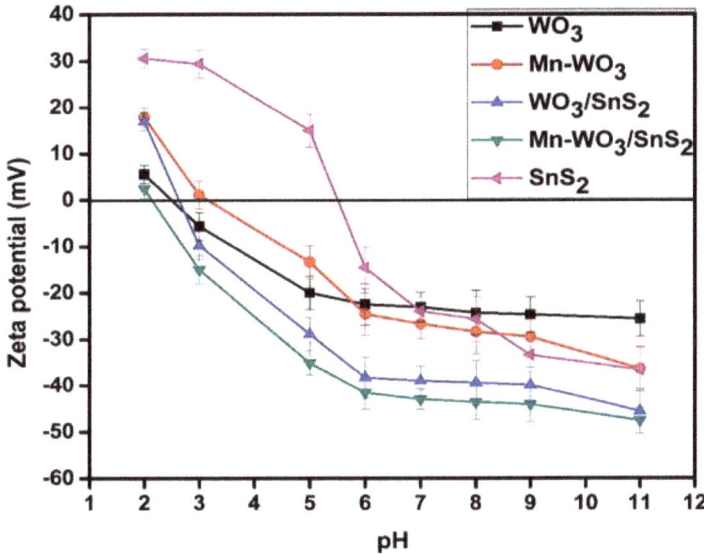

Figure 8. Surface charge of the nanoparticles.

The point of zero charge (pzc) for pristine WO_3 was observed at pH_{pzc} 2.5, which corresponds to what is reported in the literature.

A slight shift of the pzc to higher pH was observed for Mn-WO_3 (pH_{pzc} = 3.2). The shift is due to the substitution of W^{6+} by Mn^{2+} metal ions, consequently changing the overall charge of the material. Therefore, species adsorbed onto the surface of the photocatalyst change the surface charge and shift the point of zero charge of the suspended nanoparticles.

The point of zero charge for pristine SnS_2 was found to be at pH 5.5, as reported in literature. Furthermore, the heterojunction (WO_3/SnS_2) displayed a point of zero charge (2.7) at a lower pH than SnS_2 but higher than WO_3; this was attributed to synergistic effects from both counterparts (WO_3 and SnS_2) in the heterojunction.

Furthermore, introduction of Mn in the heterojunction (Mn-WO_3/SnS_2) shifted the point of zero charge to 2.1, much lower than for all the other photocatalysts.

2.7. Degradation of Chlorpyrifos

The photodegradation of chlorpyrifos using the synthesized nanoparticles is showed in Figure 9. The degradation profile for chlorpyrifos indicated an increase in removal by the nanoparticles from the WO_3 to Mn-WO_3/SnS_2 photocatalysts. The Mn-WO_3/SnS_2 nanoparticles showed high removal of chlorpyrifos due to high charge separation and lower charge impedance. Therefore, Mn-WO_3/SnS_2 represented the best performing photocatalyst with up to 95.90% chlorpyrifos removal, calculated using Equation (10).

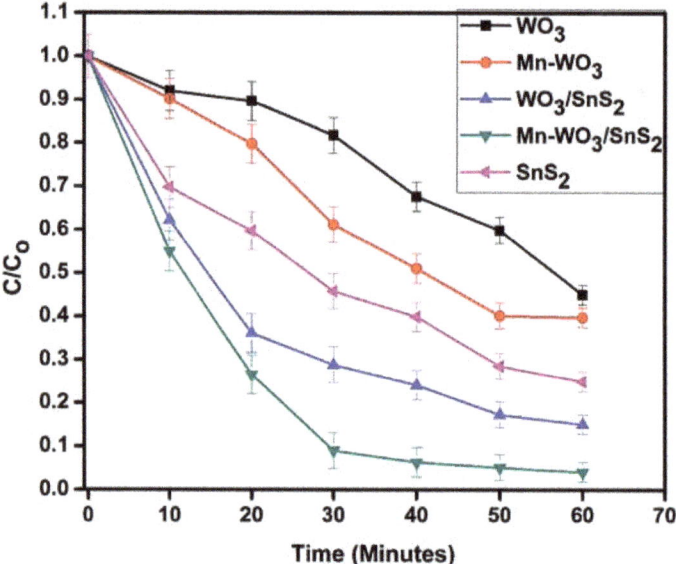

Figure 9. Degradation of chlorpyrifos (1000 ppb) using different photocatalysts at pH = 5 and 0.1 g of photocatalyst.

Figure 10 displays the percentage removal of chlorpyrifos in water within a period of 60 min. The removal efficiency for chlorpyrifos by using the nanoparticles resulted in 56.80%, 60.20%, 75.00%, 84.88% and 95.90% removal for WO_3, Mn-WO_3, SnS_2, WO_3/SnS_2 and Mn-WO_3/SnS_2, respectively.

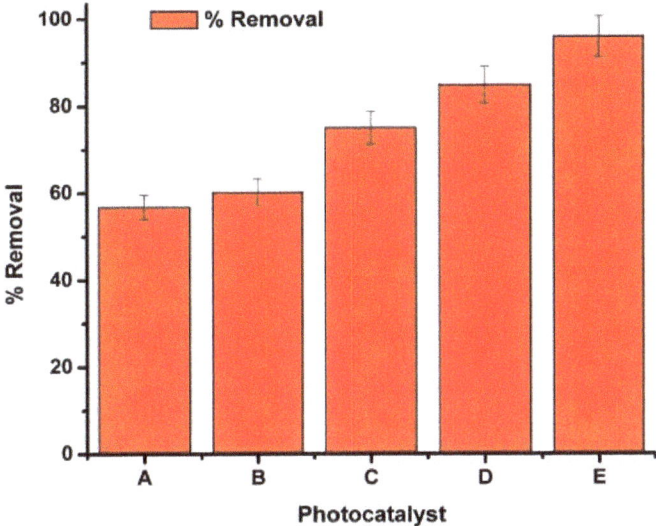

Figure 10. Percentage removal for chlorpyrifos (1000 ppb) using 0.1 g of (**A**) WO_3, (**B**) Mn-WO_3, (**C**) SnS_2, (**D**) WO_3/SnS_2 and (**E**) Mn-WO_3/SnS_2.

The reaction kinetics correspond to the percentage chlorpyrifos removal. The rate constants (K) of the reactions using the respective photocatalysts are presented (Figure 11), which were 9.3×10^{-3} $M^{-1}min^{-1}$ and 209×10^{-3} $M^{-1}min^{-1}$ for WO_3 and Mn-WO_3/SnS_2, respectively.

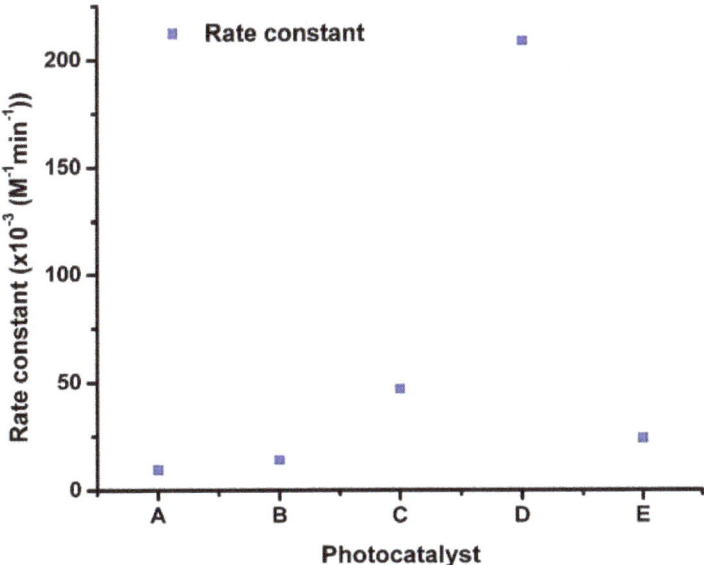

Figure 11. Rate constants of (**A**) WO_3, (**B**) Mn-WO_3, (**C**) WO_3/SnS_2, (**D**) Mn-WO_3/SnS_2 and (**E**) SnS_2.

The photodegradation reaction was fitted to Equation (10) from which the rate constant $k(M^{-1}min^{-1})$ was calculated from the gradient of the plot of 1/[C] against time (t). The reaction kinetics leading to the determination of the rate constant followed a second order reaction pathway.

The rate constants were 9.3 × 10⁻³ M⁻¹min⁻¹, 14.3 × 10⁻³ M⁻¹min⁻¹, 25.0 × 10⁻³ M⁻¹min⁻¹, 47.4 × 10⁻³ M⁻¹min⁻¹ and 209.5 × 10⁻³ M⁻¹min⁻¹, corresponding to WO_3, Mn-WO_3, SnS_2, WO_3/SnS_2 and Mn-WO_3/SnS_2, respectively (Figure 11).

The linear plot for the Mn-WO_3/SnS_2 nanoparticles kinetic studies is illustrated in Figure 12. The rate constant is 209.5 × 10⁻³ M⁻¹min⁻¹, and the R^2 is 0.9656.

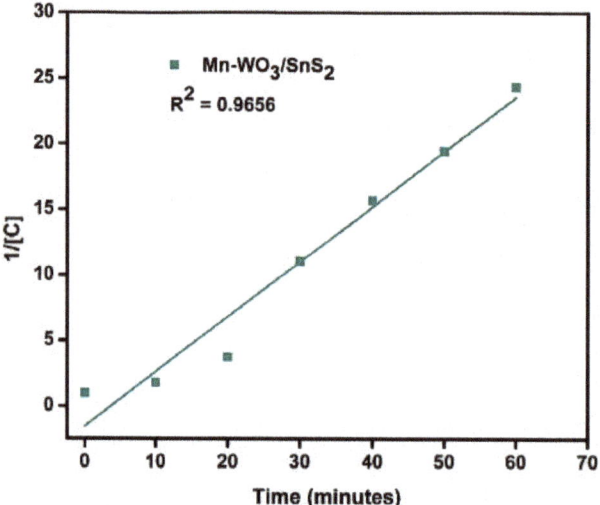

Figure 12. Photodegradation kinetics for chlorpyrifos using Mn-WO_3/SnS_2.

2.8. Effect of pH on the Photocatalytic Activity

The surface charge of the nanoparticles in a suspension is influenced by the pH of the solution. The photodegradation of chlorpyrifos using Mn-WO_3/SnS_2 nanoparticles increased with an increase in the pH of the initial solution, as illustrated in Figure 13. The point of zero charge for Mn-WO_3/SnS_2 is at pH_{pzc} = 2.13 and above that is increasingly negative, as displayed by Equations (1) and (2).

$$M\text{-}OH + H^+ = M\text{-}OH_2^+ \ (pH < pzc) \tag{1}$$

$$M\text{-}OH = M\text{-}O^- + H^+ \ (pH > pzc) \tag{2}$$

The increase in the removal was also caused by the increase in the level of deprotonation of the nanoparticles at high pH, which influences the negative charge on the surface of the photocatalyst, consequently leading to high chlorpyrifos adsorption. That was also favoured by the positive charge of chlorpyrifos in alkaline solutions from pH 5, as reported in literature. There is a transfer of holes from the inner part of the nanoparticles to the surface, whereby OH^- ions scavenge photogenerated holes and therefore yield very oxidative species such as •OH radicals. The percentage removal of chlorpyrifos achieved in 60 min was 85.6%, 94.3%, 99.8% and 99.0% at pH 3, pH 5.8, pH 7 and pH 9, respectively (Figure 13). Therefore, pH 7 was the optimum pH for chlorpyrifos removal using Mn-WO_3/SnS_2 nanoparticles and was used in the next sections. Hou et al. [31] in 2018 also reported pH 7 for optimum chlorpyrifos removal [20].

The increase in the removal was due to the increased electrostatic attraction between the photocatalyst and the chlorpyrifos that occurs when the pH is increased [28]. This causes an easy surface attachment, which implies that holes can oxidize chlorpyrifos directly and creates hydroxyl and superoxide radicals for further oxidation. As pH increased, the surface charge of the nanoparticles also became more negative, which caused increased electrostatic attraction between the nanoparticles and chlorpyrifos.

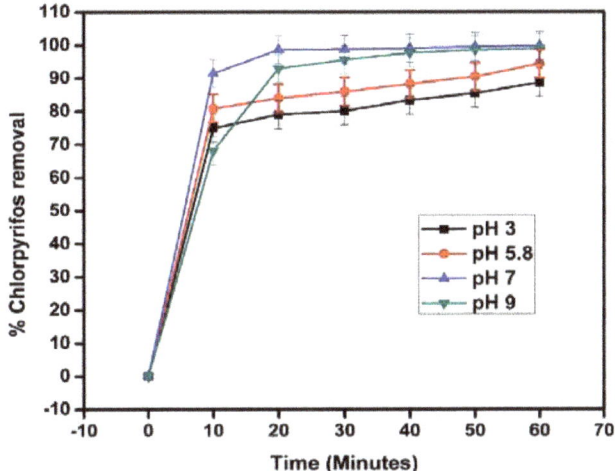

Figure 13. Degradation of 1000 ppb chlorpyrifos using 0.1 g of Mn-WO$_3$/SnS$_2$ at different pH values.

2.9. Effect of Initial Concentration

The effect of initial chlorpyrifos concentration on the photocatalytic removal was studied, and the results are shown in Figure 14. The highest removal of 99.99% was achieved at a 100 ppb chlorpyrifos concentration, followed by 99.95% at 1000 ppb, compared to 94.40%, 87.51% and 84.38% at 5 ppm, 10 ppm and 20 ppm, respectively. The concentration of 1000 ppb was chosen as the best, because it is the highest concentration for which a high percentage removal was achieved.

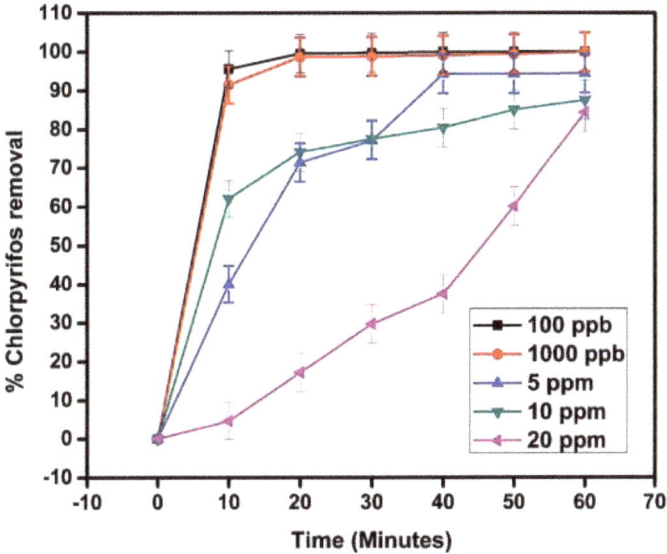

Figure 14. Effect of the initial concentration on the removal of chlorpyrifos (1000 ppb) at pH 7 using 0.1 g of Mn-WO$_3$/SnS$_2$.

The decrease in the removal of chlorpyrifos was alluded to the opacity caused by the high chlorpyrifos concentration, which prevented the photocatalyst from utilising the irradiated light to produce reactive species for degradation. Again, the high concentration scatters the light, thereby inducing screening effects [28].

2.10. Effect of Initial Photocatalyst Loading

The initial photocatalyst loading's effect on the photoactivity was studied, and the results are presented in Figure 15. The photoactivity of Mn-WO$_3$/SnS$_2$ increased when 0.5 g of photocatalyst was used, then further increased when the photocatalyst loading was 1 g. The increase is a result of increased reactive surfaces, which further increase the rate and amount of chlorpyrifos removal [28].

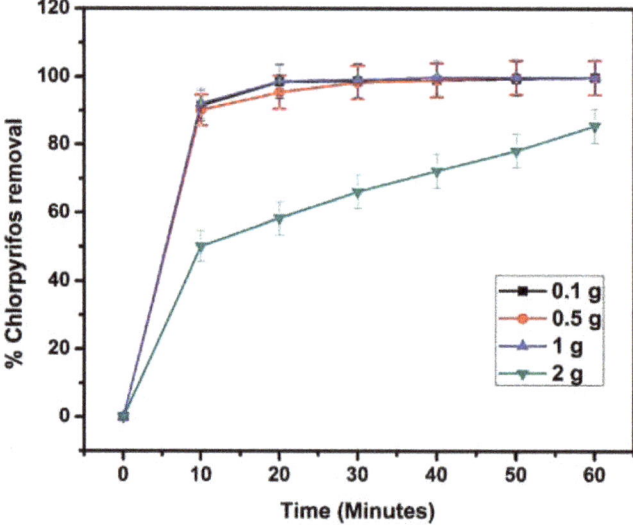

Figure 15. Effect of the initial photocatalyst loading on the photodegradation of chlorpyrifos.

A high concentration of nanoparticles results in agglomeration, which further causes light scattering and screening effects, which reduce the specific activity of the photocatalyst. This further causes opacity, which prevents the further illumination of the photocatalyst.

Therefore, a decrease in chlorpyrifos removal was observed when 2 g of Mn-WO$_3$/SnS$_2$ was used, reaching up to 85%. This is compared to 0.1 g, 0.5 g and 1 g removing up to 99.95%, 99.98% and 99.99%, respectively. Thus, 1 g was the best performing, as it reached 98% removal within 30 min of reaction time.

2.11. Mechanistic Pathway

The mechanistic and proposed degradation pathway was evaluated, and the results are shown in Figure 16. The products obtained were 3,5,6-trichloropyridin-2-ol (TCP) and O,O-dihydrogen phosphorothioite. Only the O,O-dihydrogen phosphorothioite compound and no other by-product was observed, which implies that there was a complete degradation of chlorpyrifos and TCP in the synthetic water (Figure S7).

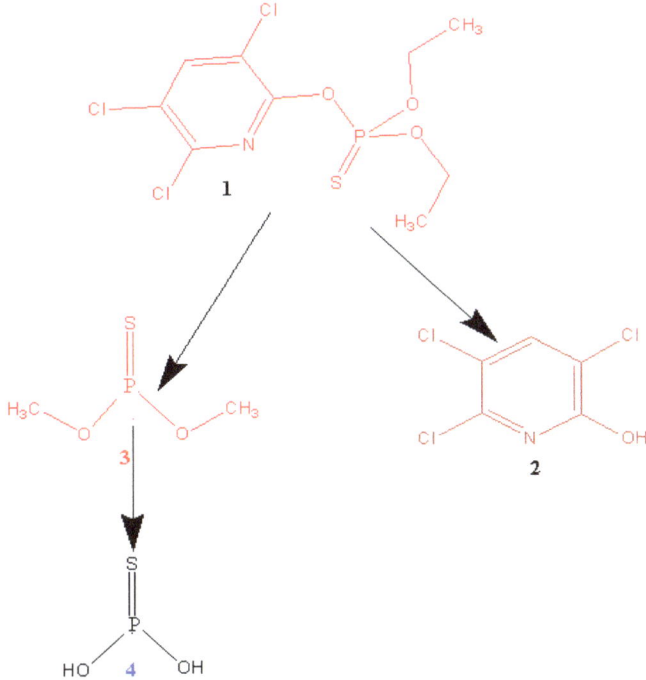

Figure 16. Proposed degradation pathway for chlorpyrifos.

3. Materials and Methods

3.1. Materials

Tungsten boride (WB) (≥97.0%), Tin(IV) chloride pentahydrate (SnCl$_4$.5H$_2$O, (98%)), manganese(II) chloride tetrahydrate (MnCl$_2$•4H$_2$O, (≥98%)), nitric acid (≥65%, Puriss), poly (vinylidene fluoride) (PVDF), N-methyl-2-pyrrolidone (NMP), silver paste, sodium sulphate (Na$_2$SO$_4$), silver/silver chloride (Ag/AgCl) electrode, sodium sulphide (Na$_2$S), chlorpyrifos (PESTNATAL, 99.9%), methanol and formic acid were all supplied by Sigma-Aldrich (Pty) Ltd., Johannesburg, South Africa. The chemicals were used as received.

Synthesis of Nanomaterial

Pristine WO$_3$ NPs were synthesised following the method developed by Xie et al. (2012) with slight modification [3]. Tungsten boride (4.12 mmol) was dissolved in 1 M HNO$_3$ (56.0 mL) under constant stirring and then transferred into a 100.0 mL Teflon-lined stainless steel autoclave. The autoclave was then sealed and placed in an oven at 190 °C for 12 h with a heating rate of 16 °C per hour; thereafter, a yellow solution was obtained. The yellow solution was further centrifuged and washed with deionised water and dried at 100 °C for 12 h in an oven, resulting in a yellow solid product of WO$_3$. The same procedure was followed to obtain Mn-WO$_3$ via the one-pot synthesis of MnCl$_2$•4H$_2$O (10.00 mmol) with WB (4.12 mmol) in 56.0 mL of HNO$_3$.

Pristine SnS$_2$ NPs were synthesised by dissolving 2.73 mmol of SnCl$_4$.5H$_2$O in 40.0 mL of deionised water under continuous stirring at 60 °C for 10 min. Thereafter, Na$_2$S (2.73 mmol) was added to the solution, and the mixture was then stirred for 10 min. The final mixture was then transferred into a 100.0 mL Teflon-lined stainless steel autoclave and heated at 180 °C for 12 h with a heating rate of 15 °C/h. The resultant solution was then centrifuged and washed with deionised water and dried at 100 °C for 12 h to obtain SnS$_2$ nanoparticles.

The heterojunction WO$_3$/SnS$_2$ was synthesised stepwise using a hydrothermal method. The first step was adapted from the method for synthesising pristine WO$_3$ and followed by the synthesis of SnS$_2$ NPs on the surface of the dispersed WO$_3$ NPs in 40.0 mL of deionised water. The SnS$_2$ synthesis was adopted from the synthesis of pristine SnS$_2$ NPs to obtain the heterojunction (WO$_3$/SnS$_2$).

Furthermore, the synthesis of the Mn-doped heterojunction WO$_3$/SnS$_2$ was carried out using hydrothermal treatment in a multistep method. Firstly, MnCl$_2$•4H$_2$O (10.0 mmol) and tungsten boride (4.12 mmol) were dissolved in 1 M HNO$_3$ (56.0 mL) under constant stirring, and thereafter, the same procedure as in the synthesis of WO$_3$ was followed to obtain a yellow solid product of Mn-WO$_3$. Furthermore, Mn-WO$_3$ (1.74 mmol) was dispersed in 40.0 mL of deionised water under continuous stirring and heating at 60 °C using a magnetic stirring hotplate. This was followed by the synthesis of SnS$_2$ NPs on the surface of Mn-WO$_3$ NPs adopted from the synthesis of pristine SnS$_2$ nanomaterials to form Mn-doped WO$_3$/SnS$_2$ heterojunction composite nanoparticles.

3.2. Characterization Techniques

The synthesised nanoparticle phases were characterized using X-ray powder diffraction (XRD) (*PANalytical X'Pert* Pro-MPD Powder Diffractometer, Almelo, Netherlands) with CuKα radiation (0.1540 nm) and a monochromator beam in a 2θ scan range from 20°–80°. The instrument power settings used were 40 kV and 40 mA with a step size of 2θ (0.0080) and a scan step time of 87.63 s. The average crystallite size was calculated using the Debye–Scherrer equation, Equation (3):

$$L = \frac{K\lambda}{\beta \cos\theta} \qquad (3)$$

where β is the full width at half maximum, λ is the X-ray wavelength (0.1541 nm) for CuKα, $K = 0.89$, and θ is the diffraction angle.

Raman spectroscopy (RamanMicro™ 200 PerkinElmer Inc., Waltham, MA, USA) with a single monochromator, a holographic notch filter and a cooled TCD, was used to detect and characterise the polymorphic forms of the NPs. The Raman spectra of the NPs were measured in a back-scattering geometry using an Ar-ion laser line (514.5 nm). Dark-field imaging was used with a power output of below 0.5 mW and an exposure time of 4.0 s. The morphological properties of the NPs were examined using high resolution transmission electron microscopy (HRTEM) (JOEL-TEM 2010) at an acceleration voltage of 200 kV. The ethanol-dispersed nanoparticles were deposited on a carbon-coated copper grid. Furthermore, selected area electron diffraction (SAED) images of the nanoparticles were captured and indexed using the CrysTBox software [29]. A field emission scanning electron microscope (FESEM) (TESCAN Vega TC instrument with VEGA 3 TESCAN software; TESCAN, Brno, Czech Republic) coupled with energy-dispersive X-ray (EDX) operated at 5.0 kV under a nitrogen gas atmosphere was used to further study the morphology and the elemental composition of the NPs. The optical properties were investigated using a UV-Vis spectrophotometer (Shimadzu UV-2450, Shimadzu Corporation, Kyoto, Japan) using diffuse reflectance spectroscopy (DRS) and BaSO$_4$ as the reference material. The band gap (E_g) of the nanomaterials and a graph of ($\alpha h\nu$) against photon energy ($h\nu$) was extrapolated following Equation (4):

$$\alpha h\nu = A(h\nu - E_g)^{n/2} \qquad (4)$$

where α is the absorption coefficient, $h\nu$ is the energy of the incident photon, A is a constant, and E_g is the band gap energy.

The value of n depends on the semiconductor transition type, which is a direct transition when n equals 0.5 and an indirect transition when n equals 2. The valence band edge potential (E_{VB}) and the conduction band edge potential (E_{CB}) were calculated using Equations (5) and (6):

$$E_{CB} = \chi - E^e - 0.5E_g \qquad (5)$$

$$E_{VB} = E_{CB} + E_g \qquad (6)$$

where E_{CB} and E_{VB} are the conduction and valance band edge potentials, respectively; χ is the electronegativity of the semiconductor (the geometric mean of the electronegativities of all the constituent atoms); E^e is the energy of free electrons on the hydrogen scale (4.5 eV); and E_g is the band gap energy of the semiconductor.

The photoluminescence spectra of the nanomaterials were obtained using a PerkinElmer fluorescence spectrometer (Model LS 45, PerkinElmer Inc., Waltham, MA, USA). A 300 W xenon lamp was used as a light source. The spectra were obtained at an excitation wavelength of 319 nm. The excitation and emission wavelengths were set at 319 nm and 605 nm, respectively. Specific surface area and pore volumes were determined using the Brunauer–Emmett–Teller (BET) method. Nitrogen was used as the adsorbate, and the nitrogen adsorption isotherms of the samples were obtained at 77K using a Micromeritics ASAP 2020 adsorption analyser (Micromeritics Instrument Corporation, Norcross, Georgia, USA). The samples were degassed before the analysis at 100 °C for 10 h. The pore volume was calculated from the amount of nitrogen adsorbed at the relative pressure (P/P_o) of 0.980.

3.3. Electrochemical Measurements

The electrochemical measurements were conducted using a potentiostat (Gamry Interface 1000 potentiostat, Gamry Instruments, Philadelphia, PA, USA) in a standard three-electrode system employing Ag/AgCl (3.0 M KCl) as the reference electrode and Pt wire as the counter electrode. The working electrodes were the prepared nanomaterial mixed with polyvinylidene fluoride (PVDF) as a binder in a 10:1 ratio respectively, dispersed in 1 mL of N-methylpyridinium (NMP) solution and ultrasonicated for 30 min to obtain a homogeneous mixture. The obtained homogeneous mixture was drop casted onto the fluorine-doped titanium oxide (FTO-glass) substrate forming a thin film. The prepared electrodes were heated at 80 °C for 12 h in air. A copper wire was thereafter attached using a silver paste for charge transfer to the potentiostat from the paste and dried in air for 24 h.

The prepared electrodes were then applied in a three-electrode system for electrochemical impedance spectroscopy (EIS) at a frequency range of 10 kHz to 0.1 Hz at an AC voltage of 10 mV rms and DC voltage of 0.45 V vs. Ag/AgCl. The current density of the working electrode was determined by running a linear sweep voltammetry scan at a scan rate of 50 mV/s. The flat-band potential (V_{fb}) values of the nanomaterials were obtained from Mott–Schottky plots (Equation (7)) at a frequency of 1000 Hz under the applied voltages of −2 to 2 V and a step voltage of 0.1 V.

$$\frac{1}{C^2} = \frac{2}{(\varepsilon\varepsilon_o A^2 e N_D)}\left[V - V_{fb} - \left(\frac{k_b T}{e}\right)\right] \qquad (7)$$

where C is the interfacial capacitance, A is the surface area of the electrode, N_D is the donor density, V is the applied potential, and V_{fb} represents the flat-band potential. The temperature with dielectric constant and permittivity of free space are represented as T, ε and ε_0, respectively. The charge of the electron (e) is 1.602×10^{-19} C, and the Boltzmann constant (k_B) is 8.617×10^{-5} eV·K^{-1}. All the electrochemical measurements were conducted in 0.1 M sodium sulphate (Na$_2$SO$_4$) solution as the electrolyte, and the values for the electrode potentials were recorded with reference to Ag/AgCl. A 300 W xenon lamp was used as the light source.

3.4. Surface Charge

Surface charge measurements were obtained using electrophoretic light scattering (ELS) with a Zetasizer NanoZS (Malvern) instrument. Zeta potential measurements were obtained using electrophoretic light scattering (ELS) to understand the surface charge of the nanomaterials as a function of the pH of the solution. The nanomaterials were suspended at 30 mg/L in deionized (DI) water. The pH of the suspensions was adjusted to a pH range of 2–10 using 1M NaOH and 1M HCl.

3.5. Degradation of Chlorpyrifos

3.5.1. Chlorpyrifos Standard Preparations

A stock solution of Chlorpyrifos (0.01 g) was prepared in 1 L of deionized water, followed by a serial dilution to make 75, 50, 25, 12.5, 6.25, and 3.125 ppb solutions. The prepared solutions were thereafter transferred into 2 mL LC-MS vials, and 1 mL of deionized water was added. The working solution was maintained at pH 5.

3.5.2. Photocatalytic Degradation of Chlorpyrifos

The photocatalytic activity of the nanomaterials was tested through the photodegradation of chlorpyrifos in synthetic water samples under visible light irradiation (Photoreactor, Lelesil Innovative Systems). The volume of the working solution was kept at 500 mL of chlorpyrifos solution. Initially, the concentration of the chlorpyrifos solution was 1 ppm and a photocatalyst loading of 0.1 g was used at pH 5. The photodegradation reaction occurred under continued magnetic stirring for 90 min under regulated temperatures of 20–25 °C, subjected to a cooling jacket using ice cubes.

The photocatalyst suspension containing chlorpyrifos was kept in the dark for 30 min before irradiation to allow equilibration. The samples were collected from the batch reaction before and after irradiation at set time intervals (10 min, 10 mL aliquots), filtered through a 0.45 µm PTFE membrane filter and transferred into a 2 mL LC-MS sample vial for analysis.

Furthermore, the optimization of reaction conditions such as the pH, initial chlorpyrifos concentration and initial photocatalyst loading were carried out. Therefore, the pH of the chlorpyrifos solution was adjusted to 3, 5, 7 and 9; the initial concentration of the pesticide (chlorpyrifos), to 100 ppb, 1 ppm, 5 ppm, 10 ppm and 20 ppm; and the photocatalyst loading, to 0.1, 0.5, 1 and 2 g.

3.6. LC-MS Measurement

Samples were analyzed using a triple quad UHPLC-MS/MS 8030 (Shimadzu Corporation) to monitor the removal of chlorpyrifos. The LC-MS/MS was fitted with a Nexera UHPLC upgrade with the capability to obtain 500 multiple reaction monitoring readings per second. The oven was equipped with a Raptor™ ARC-18 column (Restek Corporation) with a 2.7 µm pore diameter and length of 100 mm × 2.1 mm, maintained at 40 °C. The mobile phase consisted of 0.1% formic acid in water/methanol (9:1%, v/v) at a flow rate of 0.200 mL/min with a 10 µL injection volume. The ion source was electrospray ionisation (ESI) and was operated in positive mode. Meanwhile, LC-MS/MS data for the degradation intermediates were obtained after the full scan mode was run for 12 min at flow rate of 0.3 mL/min.

The percentage removal of chlorpyrifos from the synthetic water samples was calculated using Equation (8) below:

$$\% \ chlorpyrifos \ removal = \left(1 + \frac{C}{C_0}\right) \times 100 \tag{8}$$

where C_0 is the initial concentration and C is the final concentration of chlorpyrifos. The degradation products were determined by analysing the samples for a period of 60 min. The degradation pathway was then deduced from the mass/ion ratio obtained from the MS spectrum. The reaction kinetics of chlorpyrifos degradation were studied, the results were fitted to a second order model fitted, and a plot based on the calculated (1/[C]) versus reaction time was obtained following Equation (7).

$$\frac{1}{[C]_t} = kt + \frac{1}{[C]_0} \tag{9}$$

where k is the rate constant, t is time taken for the reaction, $[C]_t$ is the concentration of chlorpyrifos when time is equal to t, and $[C]_0$ is the initial concentration of chlorpyrifos. The reaction rate is thus given by Equation (10):

$$rate = k[C]^2 \tag{10}$$

4. Conclusions

The Mn-doped WO_3/SnS_2 photocatalyst was successfully synthesized, resulting in a highly crystalline structure. Rectangular rods and sheet-like shapes were observed in the composite, confirming that no shape distortion had occurred in the heterojunction photocatalyst. The composite comprises both hexagonal and monoclinic phases that correspond to SnS_2 and WO_3, respectively, as confirmed by XRD patterns and Raman spectra. As shown in the UV-Vis spectra of the composite, a shift in the *band edge* (*absorption band edge*) from the UV to the visible region (red shift) was observed in the Mn-doped WO_3/SnS_2 photocatalyst relative to that for the pristine photocatalysts. The surface area of the WO_3 was improved by more than 10 times by intrinsic doping with the Mn^{2+} ion and the formation of the heterojunction with SnS_2 to form the Mn-doped WO_3/SnS_2 photocatalyst. The Mn-doped composite was fully characterised using microscopic and spectroscopic techniques, which confirmed the synthesised composite to be Mn-WO_3/SnS_2. The Mn-doped WO_3/SnS_2 showed good electrochemical performance, ascribed to its high current density and lower interfacial charge transfer resistance, observed using electrochemical measurements (EIS), which correspond to high charge separation and a low photogenerated charge carrier recombination rate, observed using photoluminescence (PL) measurements. Chlorpyrifos has been applied extensively in agriculture, both in South Africa and other parts of the world, to fight against pests, therefore finding its way into water systems. Chlorpyrifos removal from synthetic water was investigated using Mn-WO_3/SnS_2 nanoparticles. The removal was due to the enhanced charge separation, high charge transfers and high electrostatic attraction between the nanoparticles and chlorpyrifos.

After the optimization of the reaction conditions, the chlorpyrifos removal achieved was 99.99% at pH 7 with 1 g of Mn-WO_3/SnS_2 and a 1000 ppb concentration.

The degradation pathway was also investigated, for which 3,5,6-trichloropyridin-2-ol and O,O-dihydrogen phosphorothioite were observed. Furthermore, after 60 min of the reaction, only O,O-dihydrogen phosphorothioite was detected. This implies that both chlorpyrifos and TCP were completely degraded. The results suggest that our material, Mn-WO_3/SnS_2, can completely degrade chlorpyrifos and its major degradation product.

Supplementary Materials: The following are available online at http://www.mdpi.com/2073-4344/10/6/699/s1, Table S1: Average crystallite sizes of nanomaterials; Figure S2: (a) FESEM image of pristine Mn-WO3, (b) TEM image of Mn-WO3, (c–e) elemental mapping, and (f) EDX spectrum of Mn-WO3 nanoparticles; Figure S3: (a) FESEM image of pristine SnS2, (b) TEM image of SnS2 (inset is the corresponding SAED image), (c,d) elemental mapping, and (e) EDX spectrum of SnS2 nanoparticles; Figure S4: Absorption spectra of WO3, SnS2, Mn-WO3, WO3/SnS2, and Mn-WO3/SnS2; Figure S5: EIS spectra showing the fitted spectra when obtaining the Randles circuit for (a) WO3 and Mn-WO3, and (b) SnS2, WO3/SnS2 and Mn-WO3/SnS2; Figure S6: (a–e) N2 adsorption-desorption isotherm of (a) WO3, (b) Mn-WO3, (c) Mn-WO3/SnS2, (d) WO3/SnS2, and (e) SnS2 (insets are pore volume graphs); Figure S7: Calibration curve of chlorpyrifos standards from 3.125 to 75 ppb; Figure S8: Mass spectra showing m/z ratios from 0 to 60 min; Figure S9: Fitted second order reaction kinetics graphs of the nanoparticles.

Author Contributions: Conceptualization, L.N.D., methodology, C.M.K.; validation, L.N.D., S.P.M. and C.M.K.; formal analysis, C.M.K.; investigation, C.M.K.; resources, L.N.D.; data curation, C.M.K.; writing—original draft preparation, C.M.K.; writing—review and editing, L.N.D., S.P.M.; supervision, L.N.D., S.P.M.; project administration, L.N.D.; funding acquisition, L.N.D. All authors have read and agree to the published version of the manuscript.

Funding: This research was funded by THUTHUKA NATIONAL RESEARCH FOUNDATION, grant number 15060-9119-027" and "The APC was funded by UNIVERSITY OF JOHANNESBURG-Accelerated Academic Mentorship Programme".

Acknowledgments: The authors would like to extend their gratitude to the University of Johannesburg, Faculty of Science, National Research Foundation (NRF) (TTK 15060-9119-027), TESP Eskom and the Centre for Nanomaterials Science Research, University of Johannesburg.

Conflicts of Interest: The authors declare no conflict of interest.

References

1. Dong, P.; Hou, G.; Xi, X.; Shao, R.; Dong, F. WO$_3$-based photocatalysts: Morphology control, activity enhancement and multifunctional applications. *J. Environ. Sci. Nano* **2017**, *4*, 539–557. [CrossRef]
2. Marschall, R. Semiconductor composites: Strategies for enhancing charge carrier separation to improve photocatalytic activity. *Adv. Funct. Mater.* **2014**, *24*, 2421–2440. [CrossRef]
3. Xie, Y.P.; Liu, G.; Yin, L.; Cheng, H.M. Crystal facet-dependent photocatalytic oxidation and reduction reactivity of monoclinic WO$_3$ for solar energy conversion. *J. Mater. Chem.* **2012**, *22*, 6746. [CrossRef]
4. Mahlalela, L.C.; Ngila, J.C.; Dlamini, L.N. Characterization and stability of TiO$_2$ nanoparticles in industrial dye stuff effluent. *J. Dispers. Sci. Technol.* **2017**, *38*, 584–593. [CrossRef]
5. Batzill, M. Fundamental aspects of surface engineering of transition metal oxide photocatalysts. *Int. J. Energy Environ. Sci.* **2011**, *4*, 3275. [CrossRef]
6. Harshulkhan, S.M.; Velraj, K.J.G. Structural and optical properties of Mg doped tungsten oxide prepared by microwave irradiation method. *J. Mater. Sci. Mater. Electron.* **2017**, *28*, 11794–11799. [CrossRef]
7. Parthibavarman, M.K.M.; Prabhakaran, A.K.S. One-step microwave synthesis of pure and Mn doped WO$_3$ nanoparticles and its structural, optical and electrochemical properties. *J. Mater. Sci. Mater. Electron.* **2017**, *28*, 6635–6642.
8. Lu, M.W.; Wang, Q.F.; Miao, J.; Huang, Y. Synthesis and electrochemical performances of cotton ball-like SnS$_2$ compound as anode material for lithium ion batteries. *J. Mater. Sci. Technol.* **2016**, *31*, 281–285.
9. Yamamoto, T.; Teramachi, A.; Orita, A.; Kurimoto, A.; Motoi, T.; Tanaka, T. Generation of strong acid sites on yttrium-doped tetragonal ZrO$_2$-supported tungsten oxides: Effects of dopant amounts on acidity, crystalline phase, kinds of tungsten species, and their dispersion. *J. Phys. Chem. C* **2016**, *120*, 19705–19713. [CrossRef]
10. Chae, S.Y.; Lee, C.S.; Jung, H.; Min, B.K.; Kim, J.H.; Hwang, Y.J. Insight into charge separation in WO$_3$/BiVO$_4$ heterojunction for solar water splitting. *ACS Appl. Mater. Interfaces* **2017**, *9*, 19780–19790. [CrossRef]
11. Velanganni, S.; Pravinraj, S.; Immanuel, P.; Thiruneelakandan, R. Nanostructure CdS/ZnO heterojunction configuration for photocatalytic degradation of Methylene blue. *Physics B* **2018**, *534*, 56–62. [CrossRef]
12. Afuyoni, M.; Nashed, G.; Mohammed, I. TiO$_2$ doped with SnO$_2$ and studing its structural and electrical properties. *Energy Procedia* **2011**, *6*, 11–20. [CrossRef]
13. Shaposhnik, D.; Pavelko, R.; Llobet, E.; Gispert-guirado, F.; Vilanova, X. Hydrogen sensors on the basis of SnO$_2$-TiO$_2$ systems. *Procedia Eng.* **2011**, *25*, 1133–1136. [CrossRef]
14. Tang, S.J.; Moniz, S.J.A.; Shevlin, S.A.; Martin, D.J.; Guo, Z.X.; Tang, J. Visible-light driven heterojunction photocatalysts for water splitting—A critical review. *Energy Environ. Sci.* **2015**, *8*, 731–759.
15. Simelane, S.; Ngila, J.C.; Dlamini, L.N. The effect of humic acid on the stability and aggregation kinetics of WO$_3$ nanoparticles. *Part Sci. Technol.* **2017**, *35*, 632–642. [CrossRef]
16. Georgaki, I.V.I.; Kenanakis, D.V.G.; Katsarakis, N. Synthesis of WO$_3$ catalytic powders: Evaluation of photocatalytic activity under NUV/visible light irradiation and alkaline reaction pH. *J. Sol-Gel Sci. Technol.* **2015**, *76*, 120–128.
17. Székely, I.; Kovács, G.; Baia, L.; Danciu, V.; Pap, Z. Synthesis of shape-tailored WO$_3$ micro-/nanocrystals and the photocatalytic activity of WO$_3$/TiO$_2$ composites. *Materials* **2016**, *9*, 258. [CrossRef] [PubMed]
18. Joshi, U.A.; Darwent, J.R.; Yiu, H.H.P.; Rosseinsky, M.J. The effect of platinum on the performance of WO$_3$ nanocrystal photocatalysts for the oxidation of Methyl Orange and iso-propanol. *J. Chem. Technol. Biotechnol.* **2011**, *86*, 1018–1023. [CrossRef]
19. Cai, J.; Wu, X.; Li, S.; Zheng, F. Synthesis of TiO$_2$@WO$_3$/Au nanocomposite hollow spheres with controllable size and high visible-light-driven photocatalytic activity. *ACS Sustain. Chem. Eng.* **2016**, *4*, 1581–1590. [CrossRef]
20. Dalvie, M.A.; Sosan, M.B.; Africa, A.; Cairncross, E.; London, L. Environmental monitoring of pesticide residues from farms at a neighbouring primary and pre-school in the Western Cape in South Africa. *Sci. Total Environ.* **2014**, *466–467*, 1078–1084. [CrossRef]
21. Glynnis, R.; Perry, M.; Lee, M.M.; Hoffman, E.; Delport, S.; Dalvie, M.A. Farm residence and reproductive health among boys in rural South Africa. *Environ. Int.* **2012**, *47*, 73–79.

22. Gao, J.; Naughton, S.X.; Beck, W.D.; Hernandez, C.M.; Wu, G.; Wei, Z.; Yang, X.; Bartlett, M.G.; Terry, A.V., Jr. Chlorpyrifos and chlorpyrifos oxon impair the transport of membrane bound organelles in rat cortical axons. *Neurotoxicology* **2017**, *62*, 111–123. [CrossRef]
23. Michael, J.; Madimetja, J.V.; Wepener, V. Prioritizing agricultural pesticides used in South Africa based on their environmental mobility and potential human health effects. *Environ. Int.* **2014**, *62*, 31–40.
24. Ismail, M.; Khan, H.M.; Sayed, M.; Cooper, W.J. Advanced oxidation for the treatment of chlorpyrifos in aqueous solution. *Chemosphere* **2013**, *93*, 645–651. [CrossRef] [PubMed]
25. Silvia, M.D. Highly selective sample preparation and gas chromatographic—Mass spectrometric analysis of chlorpyrifos, diazinon and their major metabolites in sludge and sludge-fertilized agricultural soils. *J. Chromatogr. A* **2006**, *1132*, 21–27.
26. Sharma, B.; Saxena, S.; Datta, A.; Arora, S. Spectrophotometric analysis of degradation of chlorpyrifos pesticide by indigenous microorganisms isolated from affected soil. *Int. J. Curr. Microb. Appl. Sci.* **2016**, *5*, 742–749. [CrossRef]
27. Chen, S.; Liu, C.; Peng, C.; Liu, H.; Hu, M.; Zhong, G. Biodegradation of Chlorpyrifos and Its Hydrolysis Product 3,5,6-Trichloro-2-Pyridinol by a New Fungal Strain *Cladosporium cladosporioides* Hu-01. *PLoS ONE* **2012**, *7*, 1–12. [CrossRef]
28. Fadaei, A.; Kargar, M. Photocatalytic degradation of chlorpyrifos in water using titanium dioxide and zinc oxide. *Fres. Environ. Bull.* **2013**, *22*, 2442–2447.
29. Ma, C.; Xu, J.; Alvarado, J.; Qu, B.; Somerville, J.; Lee, J.Y.; Meng, Y.S. Investigating the energy storage mechanism of SnS_2-rGO composite anode for advanced Na-ion batteries. *Chem. Mater.* **2015**, *27*, 5633–5640. [CrossRef]
30. Klinger, M. More features, more tools, more CrysTBox. *J. Appl. Crystallogr.* **2017**, *50*, 1–9. [CrossRef]
31. Hou, J.; Zhang, F.; Wang, P.; Wang, C.; Chen, J.; Xu, Y.; You, G.; Zhou, Q.; Li, Z. Enhanced anaerobic biological treatment of chlorpyrifos in farmland drainage with zero valent iron. *J. Chem. Eng.* **2018**, *336*, 352–360. [CrossRef]

© 2020 by the authors. Licensee MDPI, Basel, Switzerland. This article is an open access article distributed under the terms and conditions of the Creative Commons Attribution (CC BY) license (http://creativecommons.org/licenses/by/4.0/).

Article

Fast Microwave Synthesis of Gold-Doped TiO_2 Assisted by Modified Cyclodextrins for Photocatalytic Degradation of Dye and Hydrogen Production

Cécile Machut [1,*], Nicolas Kania [1], Bastien Léger [1], Frédéric Wyrwalski [1], Sébastien Noël [1], Ahmed Addad [2], Eric Monflier [1] and Anne Ponchel [1]

[1] University Artois, CNRS, Centrale Lille, Univ. Lille, UMR 8181–UCCS–Unité de Catalyse et Chimie du Solide, F-62300 Lens, France; nicolas.kania@univ-artois.fr (N.K.); bastien.leger@univ-artois.fr (B.L.); frederic.wyrwalski@univ-artois.fr (F.W.); sebastien.noel@univ-artois.fr (S.N.); eric.monflier@univ-artois.fr (E.M.); anne.ponchel@univ-artois.fr (A.P.)

[2] University Lille, CNRS, INRA, Centrale Lille, UMR 8207-UMET-Unité Matériaux et Transformations, F-59000 Lille, France; ahmed.addad@univ-lille.fr

* Correspondence: cecile.machut@univ-artois.fr

Received: 29 June 2020; Accepted: 16 July 2020; Published: 18 July 2020

Abstract: A convenient and fast microwave synthesis of gold-doped titanium dioxide materials was developed with the aid of commercially available and common cyclodextrin derivatives, acting both as reducing and stabilizing agents. Anatase titanium oxide was synthesized from titanium chloride by microwave heating without calcination. Then, the resulting titanium oxide was decorated by gold nanoparticles thanks to a microwave-assisted reduction of $HAuCl_4$ by cyclodextrin in alkaline conditions. The materials were fully characterized by UV-Vis spectroscopy, X-Ray Diffraction (XRD), Transmission Electron Microscopy (TEM), and N_2 adsorption-desorption measurements, while the metal content was determined by Inductively Coupled Plasma Optical Emission Spectroscopy (ICP-OES). The efficiency of the TiO_2@Au materials was evaluated with respect to two different photocatalytic reactions, such as dye degradation and hydrogen evolution from water.

Keywords: photocatalysis; photodegradation; nanoparticles; gold; TiO_2; cyclodextrins

1. Introduction

During the past decades, photocatalysis received extensive research interest for both limiting toxic wastes and developing clean and renewable sources of energy. Indeed, the association of a semiconductor with the sunlight in order to remove pollutants [1,2] or to produce hydrogen fuel by water splitting [3] could provide a sustainable solution to the crucial problems of environmental pollution and energy shortages [4].

Among a large number of photocatalysts, TiO_2 has been extensively investigated due to its good properties such as low cost, non-toxicity, and good stability [5,6]. Anatase phase is particularly recognized for its high photocatalytic efficiency. However, its large band gap (3.2 eV) combined with a high recombination rate of the photogenerated electron/hole pairs (e^-/h^+) reduce the photon-to-charge carriers conversion efficiency, but also limit the use in photochemical applications under visible or solar light.

In order to overcome these drawbacks and improve the photocatalytic performance of semiconductors, one of the promising strategies consists in introducing noble metals at the surface of TiO_2, such as gold nanoparticles [7]. Indeed, the combination with gold nanoparticles aims at inhibiting the electron-hole pair recombination by trapping electrons and facilitating the transfer of holes on the TiO_2 surface [8]. Gold nanoparticles (Au NPs) are also known to enhance the activity of TiO$_2$ under visible-light irradiation due to the localized surface plasmon resonance of Au NPs in the visible light spectrum [9]. However, it is well accepted that the photocatalytic activity of such TiO$_2$@Au composites can strongly depend on the particle size of Au NPs [10] and optimal synthetic conditions must be found, especially to prevent the aggregation of gold nanoparticles [11,12].

Over the last decade, cyclodextrins and derivatives have received great interest in the field of synthesis and stabilization of metallic nanoparticles in aqueous medium [13]. These macrocyclic oligosaccharides, which are well-known to form inclusion complexes with numerous guest molecules via supramolecular interactions [14], can also be used as capping agents to stabilize zerovalent metal nanoparticles, such as Au NPs. Owing to the numerous hydroxyl groups attached to the CD rims, they can also act as efficient reducing agents for the synthesis of Au NPs [15–19]. However, to the best of our knowledge, the use of cyclodextrins to prepare TiO$_2$@Au composites through simple methods of synthesis has been scarcely investigated. Most synthetic routes involve the use of chemically modified cyclodextrins bearing thiol pendant groups as metal binding sites. Their preparations require multistep and complex synthetic procedure as well as the use of time-consuming purification methods. For instance, Zhu et al. developed a method to synthesize TiO$_2$ decorated by the assembly of per-6-thio-β-cyclodextrin and gold nanoparticles. The resulting composite showed very good efficiency for the degradation of methyl orange (MO) under UV light [20]. More recently, TiO$_2$ nanosheets consisting of the combination of Au nanoparticles and mono-6-thio-β-cyclodextrin were prepared for the electrochemical detection of trace of methyl parathion pesticide [21].

Recently, we have reported a sol-gel method using cyclodextrins as both structure-directing agents and metal-complexing agents to self-assemble titania and gold colloids in composite materials with controlled porosity and uniform metal dispersion [22]. Among the various cyclodextrins examined, the TiO$_2$@Au material prepared using the commonly used randomly methylated β-CD (RAME-β-CD) have shown, after calcination, the best catalytic performance for the photodegradation of organic pollutants in water under visible light, due to a good compromise between its textural properties, crystallinity, and Au particle size. However, the preparation of such plasmonic photocatalysts involved a multistep process that occurred over several days (including acid hydrolysis, peptization, maturation, drying, and finally calcination at a high temperature of 500 °C to form Au NPs).

In recent years, microwave (MW) irradiation techniques have received considerable attention in the field of nanomaterial synthesis by inducing or enhancing chemical reactions [23–25]. The use of microwave heating may offer several advantages over conventional heating, such as shorter reaction times, higher heating rates as well as higher uniformities of the products. In the literature, a few articles were already devoted to microwave-assisted synthesis of gold nanoparticles protected by cyclodextrin derivatives [16,26,27]. As a matter of fact, Aswathy et al. synthesized β-cyclodextrin capped Au NPs with a mean diameter of 20 nm within a few minutes [16]. More recently, Stiufiuc et al. used native cyclodextrins as reducing and capping agents during the microwave reduction of the gold precursor and obtained stable monodispersed gold nanospheres covered with either α-, β- or γ-CD [26]. However, to the best of our knowledge, the stabilization and anchorage of Au NPs on titania support thanks to CDs under microwave irradiation have never been explored.

In this context, we reported hereby a novel method for elaborating TiO$_2$@Au materials from a two-step microwave-assisted synthetic route without the need for high temperature calcination. Herein, the TiO$_2$ support is synthesized using a microwave-method by hydrolysis of titanium tetrachloride while the cyclodextrins are employed afterwards to produce size-controlled gold metallic nanoparticles anchored on the support, once again under microwave irradiation. We have focused our efforts on randomly methylated-β-CD (RAME-β-CD) and 2-hydroxypropyl-β-CD (HP-β-CD), which are both highly water-soluble and readily available commercially at relatively low cost. The impact of the nature of the carbohydrate precursor is investigated and discussed on the basis of different physicochemical characterizations, including X-ray diffraction (XRD), N$_2$ adsorption-desorption analysis, transmission electron microscopy (TEM), thermogravimetry analyses (TGA), and diffuse reflectance UV-Vis spectroscopy (DRUV-Vis). Finally, the efficiency of these photocatalysts is examined with respect to two photocatalytic reactions carried out under near-UV-light irradiation (λ > 365 nm), i.e., the oxidative photodegradation of methyl orange and the hydrogen evolution reaction (HER).

2. Results and Discussion

As described in the Experimental Section, gold-doped TiO_2 materials have been synthesized at 150 °C with a fast microwave heating using cyclodextrins as reducing agent of the metal precursor and stabilizer of Au NPs. The synthetic procedure is schematically depicted in the Figure 1.

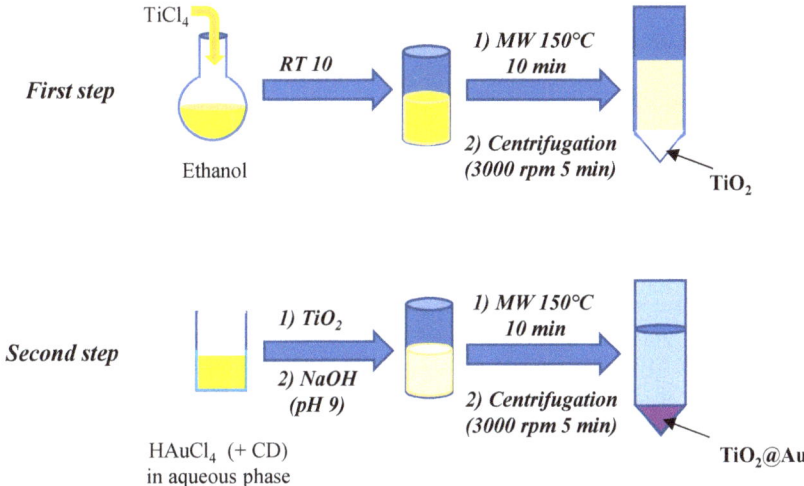

Figure 1. Schematic illustration of the two-step microwave (MW) procedure used for the TiO_2@Au materials synthesis.

Note that a bare TiO_2 control was also prepared in the same conditions as those described for the first step. These conditions were selected based on preliminary experiments, by varying the duration and power of the microwave irradiation, in order to optimize the crystallinity of our titania support. Indeed, the crystallinity is known to be a key factor in the photoactivity of TiO_2 particles. The XRD patterns of titania materials prepared from different heating programs (10, 30, and 45 min) and powers (320 and 600 W) are given for comparison in the Figure 2. With increasing the duration of heating at 320 W, we observe that the intensity of the XRD lines progressively increases and narrows, suggesting a growth in crystallite size. The planes (101), (004), (200), (105), and (211) associated to 2θ = 25.3°, 37.7°, 48°, and 55.2–55.9° respectively correspond to the anatase phase (Ti-A, JCPDS 21-1272). No XRD signals related to the presence of other crystalline phases such as rutile and brookite are detected. However, the most interesting effects are produced with the power of 600 W, which offers a very good compromise between crystallinity state and rate of anatase formation since this crystalline phase was obtained after only 10 min, this duration being considerably shorter than that applied for conventional sol-gel synthesis [28]. In line with this first optimization, the heating power of microwave irradiation was set to maximum (600 W) for all the further investigations, with a duration of temperature rise of 2 min from room temperature to 150 °C (isothermal step-time of 10 min).

The impact of the addition of gold by microwave-assisted reduction of the TiO_2 support was further investigated using mixtures of $HAuCl_4$ and modified cyclodextrins in alkaline conditions (see Figure 1, second step). We decided to use the randomly methylated β-cyclodextrin (RAME-β-CD) and the hydroxypropylated β-cyclodextrin (HP-β-CD) to stabilize Au NPs. Indeed, we particularly focused on these two CDs because of their high solubility in water and their beneficial effect on previously described gold-doped TiO_2 [22]. RAME-β-CD and HP-β-CD have also the advantages to offer a number of available hydroxyl groups (8.4 per RAME-β-CD and 21 per HP-β-CD), which are known to play an important role in the reduction processes of metal cations [16,29].

The XRD patterns of these microwave-prepared titania@Au materials are reported in the Figure 3. For comparison, the XRD pattern of a control gold-doped TiO_2 prepared in ethanol (selected as model reducing agent), but without cyclodextrin, was also included (TiO_2@Au).

It can be noticed, that in addition to the reflections of anatase (Ti-A, JCPDS 21-1272), TiO_2@Au, TiO_2@Au-RB, and TiO_2@Au-HP present broad and low intense peaks at 2θ = 38.2°, 44.2°, 64.3°, and 78.1°, which could be respectively indexed to the (111), (200), (220), and (311) planes of gold with face-centered cubic crystalline structure phase (JCPDS 04-0784). The Au crystallite sizes could have been estimated from the line broadening of the (200) diffraction peak at 2θ = 44.2° by the Debye-Sherrer equation. Interestingly, for the control-doped TiO_2 material prepared using ethanol, the size of gold crystallites is ca. 15 nm, while it significantly decreases to ca. 8–10 nm for the materials prepared with HP-β-CD and RAME-β-CD.

The textural characteristics of the titania-based materials were then evaluated by N_2 adsorption-desorption analysis. All the samples exhibit type IV adsorption isotherms with distinct hysteresis loops appearing at P/P° ≈ 0.5–0.8, thus supporting the mesoporous character of the samples with a monomodal pore size of 4 nm (Figure S1 ESI and Table 1). The specific surface areas of the bare TiO_2 (TiO_2-control) is close to those prepared by gold-doped TiO_2 (240–260 $m^2.g^{-1}$). In the same way, the pore volumes and pore size values are substantially the same whether there is gold or not. The detailed surface properties and the gold loading determined by Inductively Coupled Plasma Optical Emission Spectroscopy (ICP-OES) measurements are summarized in the Table 1. ICP-OES analysis was used to quantitatively determine the gold content in our composites. Interestingly, the gold loading (≈2 wt %) corresponds to a gold incorporation efficiency around 80–90% of the initial amount of metal used during the synthesis.

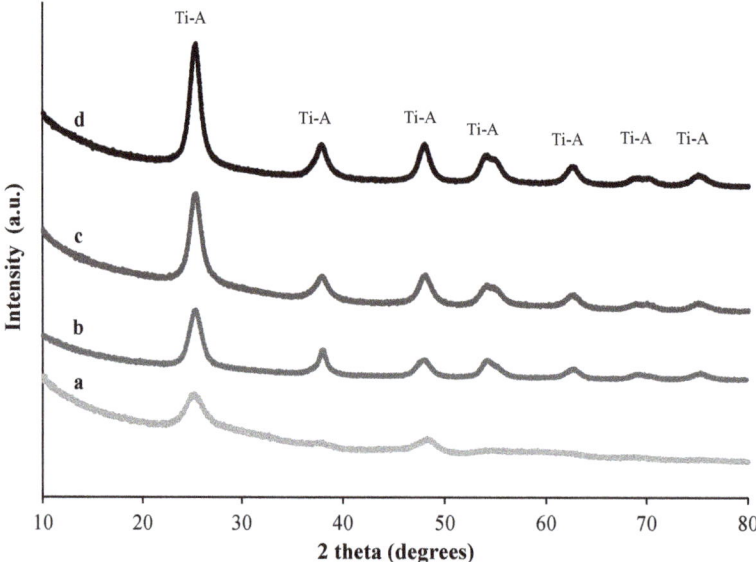

Figure 2. X-Ray Diffraction (XRD) patterns of titania-based materials prepared by microwave heating with different programs of heating (**a**) 320 W 3 min ramp then 10 min at 150 °C, (**b**) 320 W 3 min ramp then 30 min at 150 °C, (**c**) 320 W 3 min ramp then 45 min at 150 °C, and (**d**) 600 W 2 min ramp then 10 min at 150 °C.

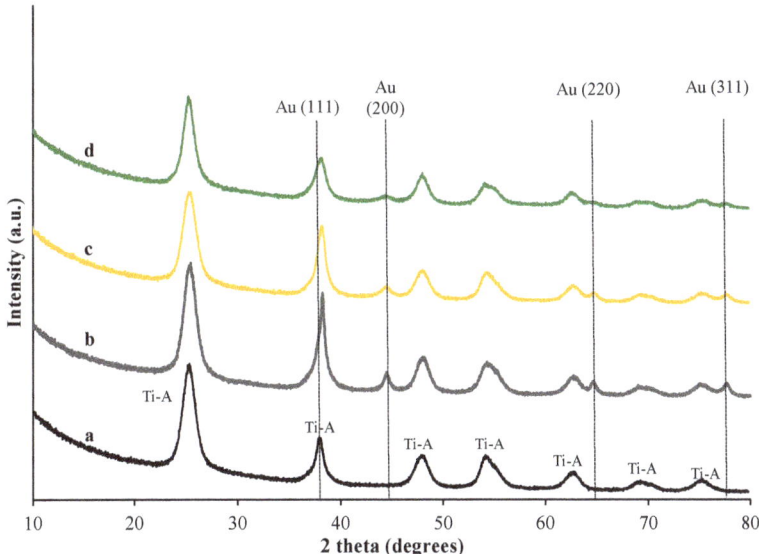

Figure 3. XRD patterns of titania materials prepared by microwave heating: (**a**) bare TiO_2, (**b**) TiO_2@Au prepared with ethanol, (**c**) TiO_2@Au-RB, and (**d**) TiO_2@Au-HP.

Table 1. Surface properties and gold loading of the different TiO_2 materials.

Sample	S_{BET} (m^2.g^{-1}) [a]	Pore Volume (cm^3. g^{-1}) [b]	Pore Size (nm) [c]	Au Content (wt %) [d]
TiO_2-control	240	0.26	4.2	-
TiO_2@Au	230	0.30	4.8	2.3
TiO_2@Au-RB	230	0.22	4.0	2.0
TiO_2@Au-HP	260	0.30	4.8	2.2

[a] Specific surface area determined by the BET (Brunauer, Emmett et Teller) method in the relative pressure range of 0.1–0.25. [b] Pore volume computed by BJH. [c] Pore size determined by BJH. [d] Gold loading determined by ICP-OES analysis.

The morphology and structure of the TiO_2@Au materials were further characterized by TEM analyses (Figure 4). Whatever the materials, the presence of Au NPs deposited onto the surface of TiO_2 is observed. When the synthesis is performed under cyclodextrin-free conditions, with ethanol as reducing agent, TEM images (Figure 4a,b) show the presence of gold nanoparticles with a mean diameter of 13.5 nm but with a relatively broad size distribution ranging from 5 to 30 nm and a standard deviation of 5.3 nm (see histogram in Figure 4c). Note that larger gold nanoparticles with diameter ranging from 44 to 78 nm can be also observed (See Figure S2 in ESI).

Although a modest decrease in the mean particle size is noticed when modified β-cyclodextrins (12.5 nm for HP-β-CD and 12.9 nm for RAME-β-CD) are introduced during the microwave-assisted synthesis, it can be seen that, for these two catalysts, gold nanoparticles are more uniformly dispersed over the TiO_2 support. Narrower size distributions with standard deviations as low as 2.5–2.8 nm (see histograms in Figure 4f,j) can be clearly found, evidencing the stabilization of small and well-dispersed spherical Au NPs, as can be seen at high magnification (See Figure S3 in ESI). It provides an intimate contact between Au NPs and the TiO_2 mesoporous support. Conversely to what was observed with the ethanol procedure, no aggregation or formation of larger particles were observed over TiO_2@Au-HP and TiO_2@Au-RB.

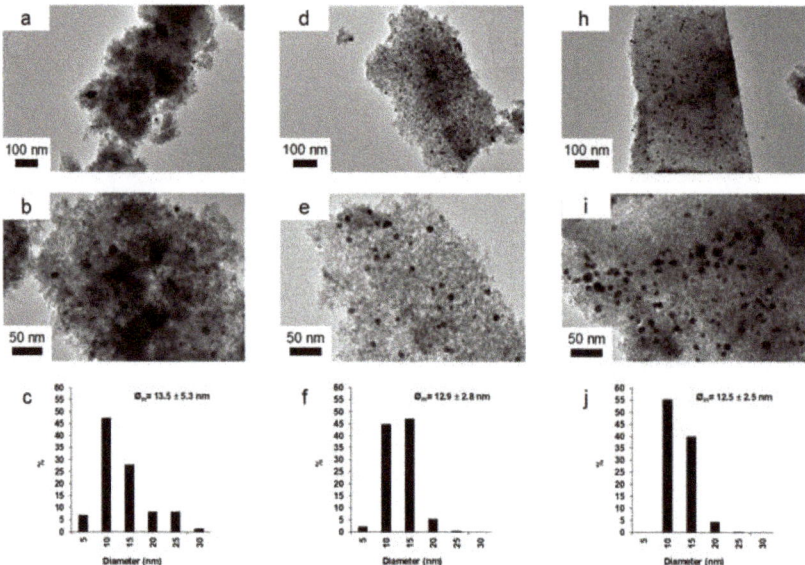

Figure 4. Transmission electron microscopy (TEM) images at magnification of ×25,000 (Scale bar = 100 nm) and ×62,000 (Scale bar = 50 nm) and size distribution of (**a**–**c**) TiO$_2$@Au, (**d**–**f**) TiO$_2$@Au-RB, (**h**–**j**) TiO$_2$@Au-HP.

As previously observed by several teams, cyclodextrins can stabilize metallic nanoparticles in aqueous solution [13]. Because of different types of interactions between the metal and the CDs (hydrophobic-hydrophobic interactions [30], non-covalent interactions between metal ions and hydroxyl groups of the CD [15]) the aggregation of gold nanoparticles can be avoided and it will result in a smaller particle size. As already observed with native CDs, RAME-β-CD and HP-β-CD are able to reduce Au^{3+} thanks to their hydroxyl groups and then interact with the gold nanoparticles in order to prevent their agglomeration [31]. To the best of our knowledge, it is the first time that modified cyclodextrins are employed as both reducing agent of gold precursor and also stabilizing agent of gold nanoparticles.

Our materials were then characterized by UV-visible diffuse reflectance spectroscopy experiment. Figure 5 shows UV–Visible absorbance spectra and Tauc plots of TiO$_2$-control, TiO$_2$@Au, TiO$_2$@Au-RB and TiO$_2$@Au-HP materials. All the titania samples exhibit a broad absorption band around 330 nm corresponding to the charge transfer from O 2p valence band to Ti 3d conduction band [32]. Thus, the large band gap energy (E$_g$) of 3.20 eV estimated for the unmodified TiO$_2$ is in agreement with typical values reported in the literature for anatase structures. However, it is worth noting that, for the gold-doped TiO$_2$ samples prepared from cyclodextrins, a slight red-shift of the absorption edge of the TiO$_2$ semiconductor toward higher wavelengths was observed compared to pure TiO$_2$. The following sequence can be established in terms of E$_g$: TiO$_2$@Au-RB (2.70 eV) < TiO$_2$@Au-HP (2.95 eV) < TiO$_2$@Au (3.20 eV) = TiO$_2$ (3.20 eV).

As previously reported, the electrons can be transferred from the excited TiO$_2$ to the metallic nanoparticles and the electron accumulation increases the Fermi level of the nanoparticle to more negative potentials. Therefore, the involved edge energy in the electron transfer from TiO$_2$ to the metallic nanoparticles is lower than bare TiO$_2$ [33]. The lowest band gap values for the TiO$_2$@Au-RB and the TiO$_2$@Au-HP materials suggest that the contact between the two inorganic phases (gold and TiO$_2$) is enhanced when cyclodextrin is used during the Au NPs synthesis and this result is in good agreement with the TEM observations. However, the smallest value was found for the TiO$_2$@Au-RB so that we can suppose that the use of the RAME-β-CD promotes the most intimate contact between the semiconductor and the metal. Further, another band is revealed at approximately 550 nm, confirming the presence of gold particles embedded in the TiO$_2$ matrix [34]. When neither cyclodextrin nor ethanol is added to the gold salt in the second step, no reduction of Au^{3+} was noticed, the resulting powder remained white and its UV-Vis spectra was similar to that obtained for the bare TiO$_2$ (see Figure S4 in ESI).

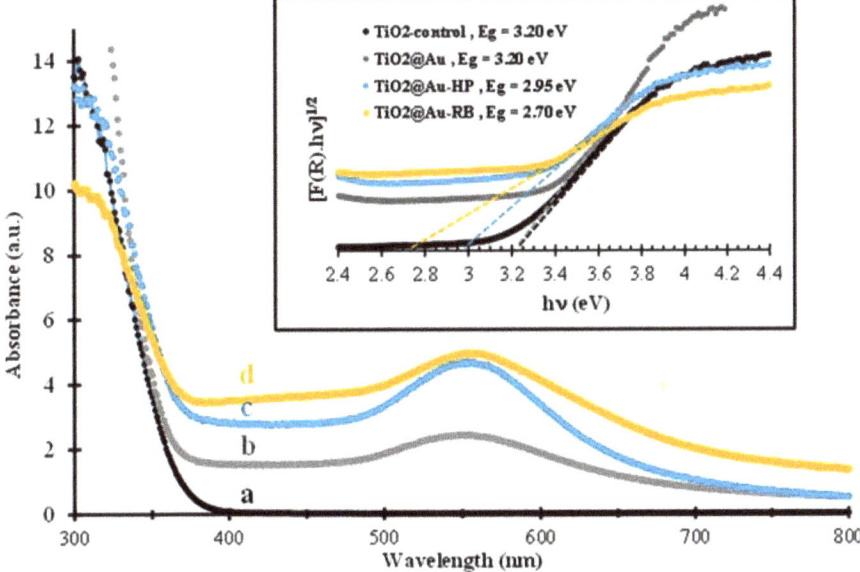

Figure 5. Diffuse Reflectance UV-Vis (DRUV-Vis) spectra of titania-based materials prepared by microwave heating: (**a**) bare TiO$_2$, (**b**) TiO$_2$@Au, (**c**) TiO$_2$@Au-HP, (**c**) and (**d**) TiO$_2$@Au-RB. In the inset, Tauc plots for the determination of the band gap values Tauc (indirect bad gap energy).

UV-Vis experiment and TEM images proved that modified CDs can act as both reducing agent of the metal precursor and capping agent of well-dispersed homogeneously dispersed Au NPs even in the presence of titanium dioxide. But to further characterize our materials and specially to know if cyclodextrins still remained in the TiO$_2$@Au-RB and the TiO$_2$@Au-HP samples, thermogravimetric analyses (TGA) were performed. The thermal profiles of TiO$_2$@Au, TiO$_2$@Au-RB, and TiO$_2$@Au-HP are shown in Figure 6.

The thermal patterns of the bare TiO$_2$ and the TiO$_2$@Au exhibit a one-step decomposition process with a weight loss in the 50–400 °C temperature range corresponding to the desorption of physically adsorbed water. The total weight loss for these samples are estimated to be 6.0 and 6.7 wt.%, respectively. The thermal profile of TiO$_2$@Au-RB exhibits a two-step decomposition process with a total weight loss of ca. 10.4 % at 1000 °C. The first weight loss (≈4%) in the 50–250 °C temperature range corresponds to the removal of physically adsorbed water, whereas the second weight loss (≈6%) in the 250–450 °C temperature range with a major weight loss at ca. 380 °C attributed to the thermal decomposition of the modified β-CD (Figure S5 in ESI). A similar profile was obtained with the TiO$_2$@Au-HP (Figure 6c) since this sample exhibited also a two-step decomposition process attributed to the removal of physically adsorbed water (≈6%) and to the thermal decomposition of residual HP-β-CD or its residues (≈11%) (see Figure S5 in ESI for the thermal profile of HP-β-CD alone). These thermal analyses proved that a small amount of saccharidic compounds (≈6 wt.% for the TiO$_2$@Au-RB and 11 wt.% for the TiO$_2$@Au-HP) remains adsorbed on our composite materials prepared with modified CDs even after the washing cycles. This result could be explained by the ability of CD derivatives to interact both with the gold nanoparticles and with the titania support. As previously described, we can suppose that after the microwave reduction, cyclodextrin derivatives could be linked to the gold nanoparticles through weak interactions and covered the outer surface of the Au NPs [16,26]. In addition, cyclodextrins are known to be able to interact with the titanium dioxide through hydrogen bounds [35,36]. In fact, the hydroxyl groups located at the exterior of the torus favored the interactions of the cyclic oligosaccharides with the surface OH groups of titania. This latter hypothesis could also explain why the amount of organic compounds is higher in the TiO$_2$@Au-HP than in the TiO$_2$@Au-RB composite: the quantity of saccharidic compounds adsorbed on titania increases with the number of hydroxyl groups of the CD [37]. Because of a higher number of hydroxyl groups (21 vs. 8.4), the HP-β-CD is more adsorbed on the titania support than the RAME-β-CD. This hypothesis could also explain the larger E_g observed for the TiO$_2$@Au-HP compared to the TiO$_2$@Au-RB composite: the residual organic compounds may reduce the contact between the Au NPs and the titanium dioxide [38].

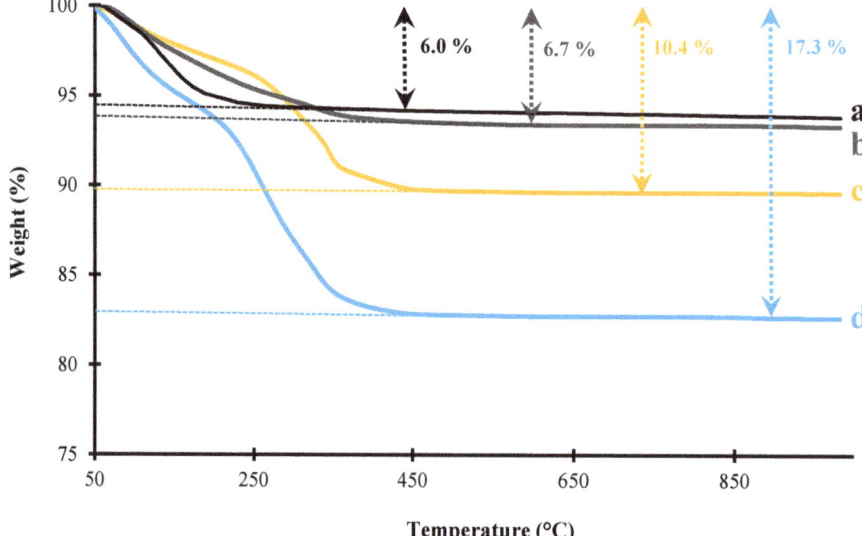

Figure 6. Thermogravimetric profiles for (**a**) bare TiO$_2$, (**b**) TiO$_2$@Au, (**c**) TiO$_2$@Au-RB, and (**d**) TiO$_2$@Au-HP.

According to the textural and structural studies, our titania-based materials exhibited interesting characteristics for photocatalytic applications. Indeed, the catalytic efficiency is known to be linked to two major physical properties: crystallinity and surface area of the photocatalysts [39].

With this microwave synthesis, only the anatase crystalline phase was obtained at low temperature (150 °C) without any additional calcination (or another thermal treatment) and this phase is known for its good activity in photocatalysis. On the other hand, good textural properties in terms of specific surface area, pore size, and pore volume could facilitate adsorption and diffusion of the target molecules onto the surface of the catalyst [40].

To confirm these hypotheses, the photocatalytic performances of the microwave gold-doped TiO$_2$ materials have been investigated through two different experiments. The redox properties of these materials have been firstly evaluated in the photodegradation of methyl orange (MO) in water. Briefly, an aqueous solution of MO (50 ppm) in the presence of the semi-conductors was irradiated at 365 nm and the concentration of the residual dye was regularly quantified by HPLC measurements. Prior to the photocatalytic study, the photostability of the organic dye was checked in a preliminary test without photocatalyst (Figure S6), and it was found that the concentration of MO remained unchanged during the 1h test period. The performances of TiO$_2$@Au-RB and TiO$_2$@Au-HP are reported in the Figure 7. For comparison, TiO$_2$@Au prepared with ethanol was also tested (Figure 7a).

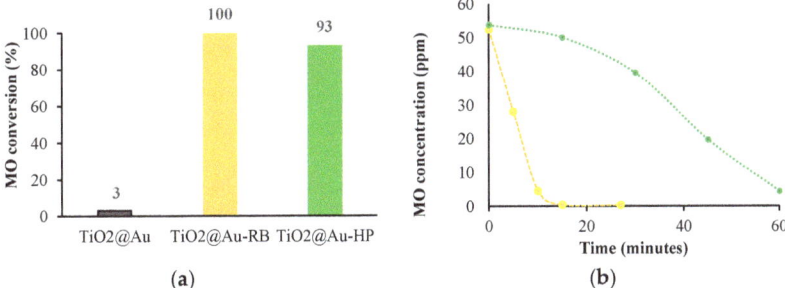

Figure 7. Photocatalytic performances of the gold-doped titania materials prepared by microwave heating for the degradation of methyl orange in near UV (λ = 365 nm): (**a**) Conversion of methyl orange after 60 min of irradiation (**b**) Evolution of the methyl orange concentration during one hour of irradiation for TiO$_2$@Au-RB (yellow) and TiO$_2$@Au-HP (green).

After 60 min under near UV irradiation, the dye was hardly degraded in the presence of the TiO$_2$@Au prepared without CD by microwave heating and this result is similar to thus obtained with bare TiO$_2$ (Figure S6). In contrast, after one hour of irradiation, the MO concentration was close to zero for the tests realized with the gold-doped TiO$_2$ prepared with modified cyclodextrins (TiO$_2$@Au-RB and TiO$_2$@Au-HP). The addition of modified CD during the synthesis of Au NPs in the presence of TiO$_2$ improved drastically the performances of the photocatalyst and this result is probably linked to the good dispersion of nanosized gold nanoparticles obtained from CDs over the support. In fact, small and well-dispersed metal islands deposited on the TiO$_2$ core are known to provide a favorable geometry for facilitating the interfacial charge transfer under UV irradiation [41]: the electrons of the titanium oxide are excited from the valence band to the conduction band and then migrate to Au clusters, which prevent the direct recombination of electrons and holes. For the TiO$_2$@Au-RB and TiO$_2$@Au-HP samples, we can suppose that the small and spherical Au NPs observed on the surface of the semi-conductor by TEM experiments act as electron sink to favor the oxidation and the reduction reactions. Conversely, large particles of metal are often harmful to the photocatalytic activity so that the TiO$_2$@Au prepared with ethanol as reducing agent was less efficient in our conditions [42]. Logically, large nanoparticles mobilize more gold atoms than small ones. With an equal metal loading, materials doped with large Au NPs offer fewer electronic reservoirs than those with small particles.

Additionally, the Figure 7b showed that the decrease of the MO amount was significantly faster in the presence of the microwave-assisted gold-doped TiO$_2$ prepared with RAME-β-CD compared to that prepared with HP-β-CD (Figure 7b). The lowest efficiency of the TiO$_2$@Au-HP compared to the TiO$_2$@Au-RB might be correlated to the highest band gap (as evidenced by DRUV-Vis experiment) and also to the amount of CDs residues in the final material (as evidenced by thermogravimetric analysis). In fact, we can suppose that the residual organic compounds decrease the contact between the semi-conductor and the gold and so reduce the electron transfer. Furthermore, the CDs residues could maybe mask some of the active sites of the semi-conductor or reduce the potential adsorption of the MO [43].

The recyclability and reuse of the most efficient photocatalyst (TiO$_2$@Au-RB) was also evaluated in the degradation of the MO. From Figure S7, it can be seen that the photocatalytic activity is stable during at least 3 runs. This study clearly showed the robustness of the catalyst and the strong embedment of the Au NPs onto the TiO$_2$ support.

Finally, we studied the behavior of the gold-doped TiO$_2$ in the production of hydrogen by photoreduction of water. Aqueous suspensions of the TiO$_2$@Au, TiO$_2$@Au-HP, and TiO$_2$@Au-RB were irradiated at 365 nm in the presence of ethanol as the sacrificial agent. The result of the amount of hydrogen produced by photoreduction of water is reported in the Figure 8a.

When the TiO$_2$@Au-HP and the TiO$_2$@Au-RB were irradiated in water, hydrogen was quickly detected and the amount of H$_2$ was quantified as about 160 and 300 µmol.h^{-1}.g^{-1} of catalyst, respectively. Compared to other TiO$_2$ catalysts in the literature [2,44], these amounts of produced hydrogen are promising since the power of our lamp is very low in comparison to Xe lamp usually used in such photocatalytic experiments. Moreover, the yield of hydrogen produced with our gold-doped catalyst was very high in comparison with that obtained with commercial anatase TiO$_2$ (<2 µmol.h^{-1}.g^{-1}) or with TiO$_2$@Au prepared with ethanol as reducing agent in the same conditions (about 3 µmol.h^{-1}.g^{-1}). As observed with the first photocatalytic test, the TiO$_2$@Au-RB was also more efficient than the TiO$_2$@Au-HP to produce hydrogen from water, probably due to the same reasons discussed above (i.e., twice as many organic compounds on the surface of the photocatalyst for the TiO$_2$@Au-HP than for the TiO$_2$@Au-RB). Finally, the amount of hydrogen produced is reproducible after several cycles of illumination (see for example Figure 8b with TiO$_2$@Au-RB) and stable during more than 10 h (ESI, Figure S8). This catalytic result proved that the introduction of small and uniform gold nanoparticles thanks to CDs reduction leads to a real boost of the photocatalytic performances of titanium dioxide even under UV irradiation and clearly confirmed the need of intimate contact with TiO$_2$ and Au to enhance the electron transfer between them.

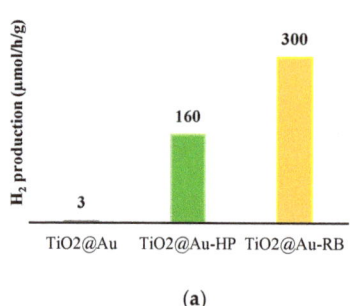
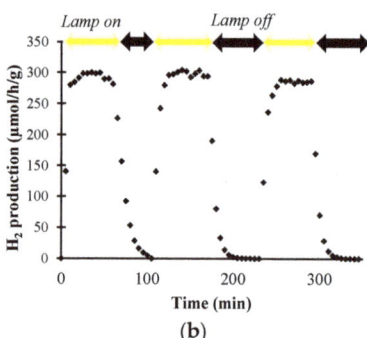

Figure 8. (**a**) Amount of hydrogen produced by photoreduction of water in the presence of gold-doped TiO$_2$ prepared by microwave heating process (100 mg of photocatalyst, 80 mL water, 20 mL ethanol, λ = 365 nm) (**b**) Evolution of the hydrogen production by photoreduction of water in the presence of TiO$_2$@Au-RB during 3 cycles of illumination.

3. Materials and Methods

3.1. Chemicals

Randomly methylated β-cyclodextrin (RAME-β-CD) with an average degree of substitution of 1.8 methyl groups per glucopyranose unit (MW 1310 g.mol^{-1}) was a gift from Wacker Chemie GmbH (Lyon, France). Hydroxypropyl-β-cyclodextrin (HP-β-CD) with an average substitution of 0.6 CH$_2$CH(OH)-CH$_3$ groups per glucopyranose unit (MW 1380 g.mol^{-1}) was purchased from Roquette (Lestrem, France). Ethanol, methyl orange (MO) and TiCl$_4$ were purchased from Sigma-Aldrich (Quentin-Fallavier, France) while HAuCl$_4$ (49 wt.%) was provided by Strem Chemicals (Bischheim, France). All these reagents were used without purification.

3.2. Preparation of the Au/TiO$_2$ Materials with Cyclodextrins

In a typical preparation, TiO$_2$ was prepared from TiCl$_4$ by microwave heating (CEM Mars instrument, Power 600 W) inspired by a method previously described by Wang et al. [45]. TiCl$_4$ (0.9 mL, 8.21 mmol) was quickly added to ethanol (25 mL) and stirred at room temperature during 10 min. Then, the yellow solution was introduced in a Teflon microwave reactor equipped with temperature and pressure probes and heated to 150 °C during 10 min. The white suspension was centrifuged at 3000 rpm during 5 min. The supernatant was evacuated and the resulting white powder of TiO_2 was added to 20 mL of an aqueous solution of $HAuCl_4$ (3.73×10^{-5} mmol) and cyclodextrin (4.04×10^{-4} mmol). Note that this CD/Au molar ratio of about 10 has been chosen to promote the synthesis of spherical gold nanoparticles, in line with a previous work reported in the literature [46]. NaOH (0.5 M) was slowly added to the solid suspension in order to adjust the pH value at about 9. Then the mixture was transferred in a Teflon microwave reactor and was finally heated under microwave irradiation with the same program used to prepare TiO_2 from $TiCl_4$ (600 W, 150 °C, 10 min). To promote the synthesis of gold nanoparticles. At the end of the heating microwave program, the suspension was centrifuged and the purple powder was thoroughly washed with water before overnight drying at 100 °C. The gold-doped TiO_2 materials synthesized by microwave heating from RAME-β-CD and HP-β-CD were named as TiO_2@Au-RB and TiO_2@Au-HP, respectively. Additionally, note that a control gold-doped TiO_2 (denoted as TiO_2@Au) was also prepared in a very similar manner as the above described procedure, by substituting cyclodextrin for ethanol during the reduction process. The syntheses and characterizations have been reproduced several times.

3.3. Characterization Methods

3.3.1. Powder X-ray Diffraction

Powder X-ray diffraction data were collected on a Siemens D5000 X-ray diffractometer (Bruker, Palaiseau, France) in a Bragg-Brentano configuration with a Cu Kα radiation source. Scans were run over the angular domains $10° < 2θ < 80°$ with a step size of 0.02° and a counting time of 2 s/step. Crystalline phases were identified by comparing the experimental diffraction patterns to Joint Committee on Powder Diffraction Standards (JCPDS) files for anatase. The treatment of the diffractograms was performed using the FullProf software [47] and its graphical interface WinPlotr [48]. The average crystallite size D was calculated from the Scherrer formula, $D = Kλ/(β \cos θ)$, where K is the shape factor (a value of 0.9 was used in this study, considering that the particles are spherical), λ is the X-ray radiation wavelength (1.54056 Å for Cu K), β is the full width at half-maximum (fwhm), and θ is the Bragg angle.

3.3.2. Nitrogen Adsorption-Desorption Isotherms

Nitrogen adsorption-desorption isotherms were collected at −196 °C using an adsorption analyzer Micromeritics Tristar 3020 (Merignac, France). Prior to analysis, 200–400 mg samples were outgassed at 100 °C overnight to remove the species adsorbed on the surface. From N_2 sorption isotherms, specific surface areas were calculated by the BET method while pore size distributions were determined using the BJH model assuming a cylindrical pore structure. The relative errors were estimated to be the following: S_{BET}, 5%; pore volume (pv) (BJH), 5%; pore size (ps) (BJH), 20%.

3.3.3. Diffuse Reflectance UV-Visible

Diffuse reflectance UV-visible spectra were collected using a Shimadzu UV-Vis NIR spectrometer (Marne-la-Vallée, France). $BaSO_4$ was used as the reference. Tauc plot analysis was performed for the calculation of the band gap energy (E_g). In fact, the E_g can be estimated by plotting $(F(R) hν)^n$ vs. $hν$ and extrapolated from linear part of the curve to the $hν$ x-axis intercept. To determine values of these forbidden energies, the absorption data were fitted to the Tauc relation for indirect band-gap transitions ($n = \frac{1}{2}$) [49].

3.3.4. Thermogravimetric Analysis (TGA) Coupled with Differential Scanning Calorimetry (DSC)

Thermogravimetric Analysis (TGA) coupled with Differential Scanning Calorimetry (DSC) analyses were performed using a Mettler Toledo TGA/DSC3+ STARe system unit (Viroflay, France).

The samples were placed in aluminum oxide crucibles of 70 µL and heated from 40 to 1000 °C at 10 °C.min^{-1} under a 50 mL.min^{-1} air flow.

3.3.5. ICP Optical Emission Spectrometry

ICP optical emission spectrometry was performed on an iCAP 7000 Thermo Scientific spectrometer (Les Ulis, France). For the quantification of gold loading, 10 mg of the Au/TiO$_2$ materials were introduced in 20 mL of aqua regia and then heated to 130 °C during one hour. Then the remaining TiO$_2$ was removed using a 0.2 µm pore filter. The resulting solution is finally diluted with pure water up to a final volume of 100 mL. The amount of gold incorporated in the material was determined using an external calibration with a gold ICP standard solution.

3.3.6. Transmission Electron Microscopy (TEM)

Transmission Electron Microscopy (TEM) bright field observations were performed on a Tecnai G2 microscope (FEI, Hillsboro, Oregon, USA) operating at an accelerating voltage of 200 kV. The Au/TiO$_2$ powder was deposited directly on a carbon coated copper grid. Metal particle size distributions have been determined from the measurement of about 200 Au NPs. The nanoparticles were found in arbitrarily chosen area of the images using the program ImageJ software.

3.4. Photocatalytic Experiments

3.4.1. Photodegradation of Methyl Orange

The photocatalytic efficiency of the titania-based materials was first evaluated in the photodegradation of methyl orange (MO) carried out using quartz reactors of 5 mL. In a typical experiment, 10 mg of photocatalyst was added to 4 mL of a solution of methyl orange (50 ppm). After 30 min in the dark, UV irradiation was performed using a led UV light lamp (Opsytec λ = 365 nm, beam size = 0.785 cm^2, power of 0.2 W.cm^{-2}). Aliquots were centrifuged at regular intervals and the MO concentration in the supernatant was determined by high-performance chromatography (HPLC, PerkinElmer, Villebon-sur-Yvette, France) analyses using a PerkinElmer Pecosphere C18 (83 mm length × 4.6 mm diameter) column. An aqueous mixture of acetonitrile (20% (v/v)) was used as the mobile phase at a flow rate of 1 mL.min^{-1}. Aliquots of 50 µL of the sample was injected and analyzed using a photodiode array detector. The MO conversion given in percentage refers to the difference in the MO concentration before irradiation (C_0) and after 1 h of irradiation (C) divided by the MO concentration before irradiation (i.e., 100 × (C_0 − C)/C_0).

3.4.2. Production of Hydrogen by Photoreduction of Water

Photocatalytic measurements for H$_2$ generation were carried out in a cylindrical pyrex reactor equipped with a quartz window by irradiating the titania-based materials in a 20 vol% ethanol-water solution (ethanol was used as hole-scavenger). As light source, we used the same LED UV light as that employed for the photocatalytic degradation of MO experiments described above in Section 3.4.1. The reactor operated at room temperature and atmospheric pressure and was kept under stirring at a constant speed of 1250 rpm. In a typical experiment, 100 mg of photocatalyst was added to a 100 mL of ethanol-water solution in the reactor. The catalytic solid suspension was then flushed with argon gas (420 mL.h^{-1}) for 60 min prior to photocatalysis. The amount of H$_2$ produced was measured on-line using a micro gas chromatograph (Micro-GC Agilent 490, Les Ulis, France) equipped with a thermal conductivity detector and two separating columns (Microsieve 10 m (5 Å) and 8 m-Paraplot U) operating with backflush injection (Ar as carrier gas).

4. Conclusions

In this work, an easy and fast preparation of Au loaded TiO$_2$ without calcination step is described. The addition of common modified cyclodextrin (methylated or hydroxypropylated) during the microwave reduction of a gold precursor in the presence of TiO$_2$ led to an efficient photocatalyst both for pollutant photodegradation and photoreduction of water under near UV irradiation. The saccharidic macrocycle was responsible for a good stabilization of gold nanoparticles in aqueous solution so that these latter could not aggregate during the microwave synthesis and were deposited uniformly on the TiO$_2$ surface. Because of its lowest number of hydroxyl groups, the RAME-β-CD seems to be less adsorbed onto the surface of the final composite after the gold reduction and represents the most promising photocatalyst. It could be now interesting to study the photocatalytic performances of our materials under solar simulated lamp. However, this new and fast synthetic approach offers promising perspectives for photocatalytic depollution process and green energy production.

Supplementary Materials: The following are available online at http://www.mdpi.com/2073-4344/10/7/801/s1, Figure S1. N2 adsorption desorption isotherms of TiO2-control (a) gold decorated titania materials prepared without CD (TiO2@Au) (b) gold decorated titania materials prepared with HP-β-CD (TiO2@Au-HP) (c) gold decorated titania materials prepared with RAME-β-CD (TiO2@Au-RB) (d), Figure S2. TEM images of TiO2@Au catalyst at magnification of ×62,000, Figure S3. TEM images of (a) TiO2@Au-RBand (b) TiO2@Au-HB at magnification of ×490,000, Figure S4. UV-vis spectra of titania materials prepared by a two-step microwave heating procedure with HAuCl4 in a second step but without CD and without ethanol, Figure S5. TGA profiles for the RAME-β-CD and the HP-β-CD, Figure S6. Evolution of methyl orange concentration under irradiation (λ = 365 nm) as a function of time in the absence (open circle) or presence of the bare TiO2 prepared by microwave process (filled circle). Reaction conditions: TiO2, m = 10 mg; methyl orange solution, V = 4 mL (50 ppm) Figure S7. Performance of TiO2@Au-RB in three consecutive tests with reuse of the catalyst. Reaction conditions: 4 mL of a solution of methylorange (50 ppm), 10 mg of TiO2@Au-RB (λ = 365 nm, t = 10 min), Figure S8. Production of hydrogen by photoreduction of water (80 mL) in the presence of TiO2@Au-RB (100 mg) and ethanol (20 mL) as sacrificial agent (λ = 365 nm).

Author Contributions: Synthesis, catalytic tests, ICP, and UV experiments, C.M.; N$_2$ adsorption-desorption measurements and IR spectroscopy, N.K.; TEM analysis, B.L. and A.A.; XRD, A.P. and F.W.; catalysis, S.N.; supervision and reviewing results, reviewing the manuscript, English writing, C.M., E.M., and A.P. All authors have read and agreed to the published version of the manuscript.

Funding: This research received no external funding.

Acknowledgments: The TEM facility in Lille (France) is supported by the Conseil Régional du Nord Pas de Calais and the European Regional Development Fund (ERF). Chevreul Institute (FR 2638), Ministère de l'Enseignement Supérieur, de la Recherche et de l'Innovation, Région Hauts-de-France and FEDER are acknowledged for supporting and funding partially this work. The authors are grateful to the University of Artois for supporting this research through the Quality Research Bonus (Micro-GC Agilent 490 in 2018).

Conflicts of Interest: The authors declare no conflict of interest.

References

1. Youa, J.; Guoa, Y.; Guob, R.; Liub, X. A review of visible light-active photocatalysts for water disinfection: Features and prospects. *Chem. Eng. J.* **2019**, *373*, 624–641. [CrossRef]
2. Boyjoo, Y.; Sun, H.; Liu, J.; Pareek, V.K.; Wang, S. A review on photocatalysis for air treatment: From catalyst development to reactor design. *Chem. Eng. J.* **2017**, *310*, 537–559. [CrossRef]
3. Jafari, T.; Moharreri, E.; Amin, A.S.; Miao, R.; Song, W.; Suib, S.L. Photocatalytic Water Splitting—The Untamed Dream: A Review of Recent Advances. *Molecules* **2016**, *21*, 900–929. [CrossRef]
4. Daghrir, R.; Drogui, P.; Robert, D. Modified TiO$_2$ for Environmental Photocatalytic Applications: A Review. *Ind. Eng. Chem. Res.* **2013**, *52*, 3581–3599. [CrossRef]
5. Chen, X.; Mao, S.S. Titanium Dioxide Nanomaterials: Synthesis, Properties, Modifications and Applications. *Chem. Rev.* **2007**, *107*, 2891–2959. [CrossRef]
6. Haider, A.J.; Jameel, Z.N.; Al-Hussaini, I.H. Review on: Titanium Dioxide Applications. *Energy Procedia* **2019**, *157*, 17–29. [CrossRef]
7. Khaki, M.R.D.; Shafeeyan, M.S.; Raman, A.A.A.; Daud, W.M.A.W. Application of doped photocatalysts for organic pollutant degradation—A review. *J. Environ. Manag.* **2017**, *198*, 78–94. [CrossRef] [PubMed]
8. Truppi, A.; Petronella, F.; Placido, T.; Margiotta, V.; Lasorella, G.; Giotta, L.; Giannini, C.; Sibillano, T.; Murgolo, S.; Mascolo, G.; et al. Gram-scale synthesis of UV-vis light active plasmonic photocatalytic nanocomposite based on TiO$_2$/Au nanorods for degradation of pollutants in water. *Appl. Catal. B Environ.* **2019**, *243*, 604–613. [CrossRef]

9. Cheng, L.; Zhang, D.; Liao, Y.; Li, F.; Zhang, H.; Xiang, Q. Constructing functionalized plasmonic gold/titanium dioxide nanosheets with small gold nanoparticles for efficient photocatalytic hydrogen evolution. *J. Colloid Interface Sci.* **2019**, *555*, 94–103. [CrossRef] [PubMed]
10. Iliev, V.; Tomova, D.; Bilyarska, L.; Tyuliev, G. Influence of the size of gold nanoparticles deposited on TiO_2 upon the photocatalytic destruction of oxalic acid. *J. Mol. Catal. A Chem.* **2007**, *263*, 32–38. [CrossRef]
11. Zhou, H.; Zheng, L.; Jia, H. Facile control of the self-assembly of gold nanoparticles by changing the capping agent structures. *Colloids Surf. A Physicochem. Eng. Asp.* **2014**, *450*, 9–14. [CrossRef]
12. Zielińska-Jurek, A.; Kowalska, E.; Sobczak, J.W.; Lisowski, W.; Ohtani, B.; Zaleska, A. Preparation and characterization of monometallic (Au) and bimetallic (Ag/Au) modified-titania photocatalysts activated by visible light. *Appl. Catal. B Environ.* **2011**, *101*, 504–514. [CrossRef]
13. Noël, S.; Léger, B.; Ponchel, A.; Philippot, K.; Denicourt-Nowicki, A.; Roucoux, A.; Monflier, E. Cyclodextrin-based systems for the stabilization of metallic(0) nanoparticles and their versatile applications in catalysis. *Catal. Today* **2014**, *235*, 20–32. [CrossRef]
14. Connors, K.A. The Stability of Cyclodextrin Complexes in Solution. *Chem. Rev.* **1997**, *97*, 1325–1358. [CrossRef] [PubMed]
15. Bhoi, V.I.; Kumar, S.; Murthy, C.N. Cyclodextrin encapsulated monometallic and inverted core–shell bimetallic nanoparticles as efficient free radical scavengers. *New J. Chem.* **2016**, *40*, 1396–1402. [CrossRef]
16. Aswathy, B.; Avadhani, G.S.; Suji, S.; Sony, G. Synthesis of β-cyclodextrin functionalized gold nanoparticles for the selective detection of Pb^{2+} ions from aqueous solution. *Front. Mater. Sci.* **2012**, *6*, 168–175. [CrossRef]
17. Woo Chung, J.; Guo, Y.; Kwak, S.-Y.; Priestley, R.D. Understanding and controlling gold nanoparticle formation from a robust self-assembled cyclodextrin solid template. *J. Mater. Chem.* **2012**, *22*, 6017–6026. [CrossRef]
18. Huang, T.; Meng, F.; Qi, L. Facile Synthesis and One-Dimensional Assembly of Cyclodextrin-Capped Gold Nanoparticles and Their Applications in Catalysis and Surface-Enhanced Raman Scattering. *J. Phys. Chem. C* **2009**, *113*, 13636–13642. [CrossRef]
19. Pande, S.; Ghosh, S.K.; Praharaj, S.; Panigrahi, S.; Basu, S.; Jana, S.; Pal, A.I.; Tsukuda, T.; Pal, T. Synthesis of Normal and Inverted Gold-Silver Core-Shell Architectures in α-Cyclodextrin and Their Applications in SERS. *J. Phys. Chem. C* **2007**, *111*, 10806–10813. [CrossRef]
20. Zhu, H.; Goswami, N.; Yao, Q.; Chen, T.; Liu, Y.; Xu, Q.; Chen, D.; Lu, J.; Xie, J. Cyclodextrin–gold nanocluster decorated TiO_2 enhances photocatalytic decomposition of organic pollutants. *J. Mater. Chem. A* **2018**, *6*, 1102–1108. [CrossRef]
21. Fu, X.-C.; Zhang, C.; Li, X.-H.; Zhang, J.; Wei, G. Mono-6-thio-β-cyclodextrin-functionalized AuNP/two-dimensional TiO_2 nanosheet nanocomposite for the electrochemical determination of trace methyl parathion in water. *Anal. Methods* **2019**, *11*, 4751–4760. [CrossRef]
22. Lannoy, A.; Bleta, R.; Machut-Binkowski, C.; Addad, A.; Monflier, E.; Ponchel, A. Cyclodextrin-Directed Synthesis of Gold-Modified TiO_2 Materials and Evaluation of Their Photocatalytic Activity in the Removal of a Pesticide from Water: Effect of Porosity and Particle Size. *ACS Sustain. Chem. Eng.* **2017**, *5*, 3623–3630. [CrossRef]
23. Wen, P.; Wu, Z.; Han, Y.; Cravotto, G.; Wang, J.; Ye, B.-C. Microwave-Assisted Synthesis of a Novel Biochar-Based Slow-Release Nitrogen Fertilizer with Enhanced Water-Retention Capacity. *ACS Sustain. Chem. Eng.* **2017**, *5*, 7374–7382. [CrossRef]
24. Martina, K.; Baricco, F.; Berlier, G.; Caporaso, M.; Cravotto, G. Efficient Green Protocols for Preparation of Highly Functionalized β-Cyclodextrin-Grafted Silica. *ACS Sustain. Chem. Eng.* **2014**, *2*, 2595–2603. [CrossRef]
25. Tabasso, S.; Calcio Gaudino, E.; Acciardo, E.; Manzoli, M.; Bonelli, B.; Cravotto, G. Microwave-Assisted Protocol for Green Functionalization of Thiophenes With a Pd/b-Cyclodextrin Cross-Linked Nanocatalyst. *Front. Chem.* **2020**, *8*, 253. [CrossRef]
26. Stiufiuc, G.; Toma, V.; Moldovan, I.; Stiufiuc, R.; Lucaciu, C.M. One pot microwave assisted synthesis of cyclodextrins capped spherical gold nanoparticles. *Dig. J. Nanomater. Biostruct.* **2017**, *12*, 1089–1095.
27. Tsuji, M. Microwave-Assisted Synthesis of Metallic Nanomaterials in Liquid Phase. *ChemistrySelect* **2017**, *2*, 805–819. [CrossRef]

28. Qi, K.; Xin, J.H. Room-Temperature Synthesis of Single-Phase Anatase TiO$_2$ by Aging and its Self-Cleaning Properties. *ACS Appl. Mater. Interfaces* **2010**, *2*, 3479–3485. [CrossRef]
29. Pestovsky, Y.S.; Martínez-Antonio, A. Synthesis of Gold Nanoparticles by Tetrachloroaurate Reduction with Cyclodextrins. *Quim. Nova* **2018**, *41*, 926–932.
30. Liu, Y.; Male, K.B.; Bouvrette, P.; Luong, J.H.T. Control of the Size and Distribution of Gold Nanoparticles by Unmodified Cyclodextrins. *Chem. Mater.* **2003**, *15*, 4172–4180. [CrossRef]
31. Zhao, Y.; Huang, Y.; Zhu, H.; Zhu, Q.; Xia, Y. Three-in-One: Sensing; Self-Assembly and Cascade Catalysis of Cyclodextrin Modified Gold Nanoparticles. *J. Am. Chem. Soc.* **2016**, *138*, 16645–16654. [CrossRef] [PubMed]
32. Morawa Eblagon, K.; Pastrana-Martínez, L.M.; Pereira MF, R.; Figueiredo, J.L. Cascade conversion of cellobiose to gluconic acid: The large impact of the small modification of electronic interaction on the performance of Au/TiO$_2$ bifunctional catalysts. *Energy Technol.* **2018**, *6*, 1675–1686. [CrossRef]
33. Oros-Ruiza, S.; Zanellaa, R.; Prado, B. Photocatalytic degradation of trimethoprim by metallic nanoparticles supported on TiO$_2$-P25. *J. Hazard. Mater.* **2013**, *263*, 28–35. [CrossRef] [PubMed]
34. Wang, X.; Mitchell, D.R.G.; Prince, K.; Atanacio, A.J.; Caruso, R.A. Gold nanoparticle incorporation into porous titania networks using agarose gel templating technique for photocatlytic applications. *Chem. Mater.* **2008**, *20*, 3917–3926. [CrossRef]
35. Willner, I.; Eichen, Y.; Willner, B. Supramolecular semiconductor receptor assemblies: Improved electron transfer at TiO$_2$-β-cyclodextrin colloid interfaces. *Res. Chem. Intermed.* **1994**, *20*, 681–700. [CrossRef]
36. Zhang, X.; Wu, F.; Wang, Z.; Guo, Y.; Deng, N. Photocatalytic degradation of 4;4′-biphenol in TiO$_2$ suspension in the presence of cyclodextrins: A trinity integrated mechanism. *J. Mol. Catal. A Chem.* **2009**, *301*, 134–139. [CrossRef]
37. Lannoy, A.; Kania, N.; Bleta, R.; Fourmentin, S.; Machut-Binkowski, C.; Monflier, E.; Ponchel, A. Photocatalysis of Volatile Organic Compounds in water: Towards a deeper understanding of the role of cyclodextrins in the photodegradation of toluene over titanium dioxide. *J. Colloid Interface Sci.* **2016**, *461*, 317–325. [CrossRef]
38. May-Masnou, A.; Soler, L.; Torras, M.; Salles, P.; Llorca, J.; Roig, A. Fast and Simple Microwave Synthesis of TiO$_2$/Au Nanoparticles for Gas-Phase Photocatalytic Hydrogen Generation. *Front. Chem.* **2018**, *10*, 110. [CrossRef]
39. Kanna, M.; Wognawa, S. Mixed amorphous and nanocrystalline TiO$_2$ powders prepared by sol-gel method: Characterization and photocatalytic study. *Mater. Chem. Phys.* **2008**, *110*, 166–175. [CrossRef]
40. Wang, X.; Caruso, R.A. Enhancing photocatalytic activity of titania materials by using porous structures and the addition of gold nanoparticles. *J. Mater. Chem.* **2011**, *21*, 20–28. [CrossRef]
41. Dawson, A.; Kamat, P.V. Semiconductor-Metal Nanocomposites. Photoinduced Fusion and Photocatalysis of Gold-Capped TiO$_2$ (TiO$_2$/Gold) Nanoparticles. *J. Phys. Chem. B* **2001**, *105*, 960–966. [CrossRef]
42. Gołąbiewska, A.; Malankowska, A.; Jarek, M.; Lisowski, W.; Nowaczyk, G.; Jurga, S.; Zaleska-Medynska, A. The effect of gold shape and size on the properties and visible light-induced photoactivity of Au-TiO$_2$. *Appl. Catal. B Environ.* **2016**, *196*, 27–40. [CrossRef]
43. Luna, A.L.; Matter, F.; Schreck, M.; Wohlwend, J.; Tervoort, E.; Colbeau-Justin, C.; Niederberger, M. Monolithic metal-containing TiO$_2$ aerogels assembled from crystalline pre-formed nanoparticles as efficient photocatalysts for H$_2$ generation. *Appl. Catal. B Environ.* **2020**, *267*, 118660. [CrossRef]
44. Wang, B.; Shen, S.; Mao, S.S. Black TiO$_2$ for solar hydrogen conversion. *J. Mater.* **2017**, *3*, 96–111.
45. Wang, X.; Tian, J.; Fei, C.; Lv, L.; Wang, Y.; Cao, G. Rapid construction of TiO$_2$ aggregates using microwave assisted synthesis and its application for dye-sensitized solar cells. *RSC Adv.* **2015**, *5*, 8622–8629. [CrossRef]
46. Huang, Y.; Li, D.; Li, J. β-Cyclodextrin controlled assembling nanostructures from gold nanoparticles to gold nanowires. *Chem. Phys. Lett.* **2004**, *389*, 14–18. [CrossRef]
47. Rodriguez-Carvajal, J. FULLPROF: A Program for Rietveld Refinement and Pattern Matching Analysis. In *Abstracts of the Satellite Meeting on Powder Diffraction of the XV Congress of the IUCR, Toulouse, France*; International Union of Crystallography: Chester, UK, 1990; p. 127.

48. Roisnel, T.; Rodriguez-Carvajal, J. WinPLOTR: A Windows Tool for Powder Diffraction Pattern Analysis. In *Materials Science Forum, Proceedings of the 7th European Powder Diffraction Conference (EPDIC 7), Barcelona, Spain, 20–23 May 2000*; Delhez, R., Mittenmeijer, E.J., Eds.; Trans Tech Publications: Zurich, Switzerland, 2000; pp. 118–123.
49. Khore, S.K.; Kadam, S.R.; Naik, S.D.; Kale, B.B.; Sonawane, R.S. Solar light active plasmonic Au@TiO_2 nanocomposite with superior photocatalytic performance for H_2 production and pollutant degradation. *New J. Chem.* **2018**, *42*, 10958–10968. [CrossRef]

© 2020 by the authors. Licensee MDPI, Basel, Switzerland. This article is an open access article distributed under the terms and conditions of the Creative Commons Attribution (CC BY) license (http://creativecommons.org/licenses/by/4.0/).

Article

Effect of Potential and Chlorides on Photoelectrochemical Removal of Diethyl Phthalate from Water

Laura Mais, Simonetta Palmas, Michele Mascia and Annalisa Vacca *

Dipartimento di Ingegneria Meccanica, Chimica e dei Materiali, Università degli Studi di Cagliari, Via Marengo 2, 09123 Cagliari, Italy; laura.mais@unica.it (L.M.); simonetta.palmas@dimcm.unica.it (S.P.); michele.mascia@unica.it (M.M.)
* Correspondence: annalisa.vacca@dimcm.unica.it

Abstract: Removal of persistent pollutants from water by photoelectrocatalysis has emerged as a promising powerful process. Applied potential plays a key role in the photocatalytic activity of the semi-conductor as well as the possible presence of chloride ions in the solution. This work aims to investigate these effects on the photoelectrocatalytic oxidation of diethyl phthalate (DEP) by using TiO_2 nanotubular anodes under solar light irradiation. PEC tests were performed at constant potentials under different concentration of NaCl. The process is able to remove DEP following a pseudo-first order kinetics: values of k_{app} of 1.25×10^{-3} min^{-1} and 1.56×10^{-4} min^{-1} have been obtained at applied potentials of 1.8 and 0.2 V, respectively. Results showed that, depending on the applied potential, the presence of chloride ions in the solution affects the degradation rate resulting in a negative effect: the presence of 500 mM of Cl$^-$ reduces the value of k_{app} by 50 and 80% at 0.2 and 1.8 V respectively.

Keywords: diethyl phthalate; photoelectrochemical degradation; persistent organic pollutants; chloride ions; TiO_2 nanotubes

Citation: Mais, L.; Palmas, S.; Mascia, M.; Vacca, A. Effect of Potential and Chlorides on Photoelectrochemical Removal of Diethyl Phthalate from Water. *Catalysts* **2021**, *11*, 882. https://doi.org/10.3390/catal11080882

Academic Editor: Bruno Fabre

Received: 17 June 2021
Accepted: 19 July 2021
Published: 22 July 2021

Publisher's Note: MDPI stays neutral with regard to jurisdictional claims in published maps and institutional affiliations.

Copyright: © 2021 by the authors. Licensee MDPI, Basel, Switzerland. This article is an open access article distributed under the terms and conditions of the Creative Commons Attribution (CC BY) license (https://creativecommons.org/licenses/by/4.0/).

1. Introduction

The application of photoelectrochemical process for polluted waters and wastewaters has been gaining more and more attention thanks to the possibility to obtain electrical energy from renewable energy sources, rather than from fossil fuels [1]. The technique exploits the synergy between photochemistry and electrochemistry: from one side, the photochemical process increases its efficiency as the bias potential lowers recombination of the photogenerated charges, from the other side the photo-potential generated on the semiconductor depolarizes the cell improving the yield of the electrochemical process [2].

Considering the application to real matrices, the effect of the composition of the water to be treated plays a crucial role, with particular regard to the presence of chlorides, which are ubiquitous ions in water and wastewater. Several studies on the photochemical process using TiO_2 highlighted a negative effect of the presence of chloride: the inhibiting effect has been ascribed both to the competitive adsorption between the pollutant molecules and Cl$^-$ towards the surface-active sites of TiO_2, or to the scavenging function of chloride ions towards holes and hydroxyl radicals [3,4]. Piscopo et al. [5] showed different effects on the degradation rate of two pollutants depending on the chloride concentration, the nature of the organics and the pH: in the case of poorly adsorbed molecules, if the pH favored the adsorption of Cl$^-$, even low concentration of chloride strongly affected the degradation.

Several papers evidenced the key role of pH in the photocatalytic degradation using TiO_2: point of zero charge (pH$_{pzc}$) plays a crucial role in determining the surface charge of photocatalyst and, in turn, its interaction with charged molecules or ions. When the pH is higher than the pH$_{pzc}$, the polarity of TiO_2 surface is negative and the electrostatic repulsion toward anionic compounds dominates [6–8]. Moreover, since hydroxyl radicals can be formed by the reaction between hydroxide ions and positive holes, the hydroxyl radicals are considered as the predominant species at neutral or high pH, while at low pH the holes are considered the major oxidizing species [9]. Regarding the scavenging effect, chloride can react with HO• radicals and holes, allowing the formation of less reactive chloride radical (Cl•) and dichloride radicals ($Cl_2\bullet^-$) [10–12]: the oxidized chloride may also recombine with photogenerated electrons quenching the photogenerated charge carriers [13].

Different considerations may be made when photoelectrocatalysis is considered: in this case, heterogenous photocatalysis can be improved by the application of a bias potential to obtain a more effective separation of photogenerated charges, thereby increasing the lifetime of electron–hole pairs. In the photoelectrocatalytic process, the increases of the applied potential can accelerate the photogenerated electrons toward the external circuit, generating the bending of the conduction and valence bands, with the consequent formation of a space charge layer. Thus, the recombination of the e^-/h^+ pairs may be decreased or totally prevented, improving the photocatalytic performance [14,15]. Moreover, increase in the potential can empty the defects where the photogenerated charges are trapped, enhancing the photoactivity [16].

The presence of chloride in a photoelectrochemical process exerts a different effect with respect to the photochemical one: in fact, unlike the inhibitory effects found in photocatalysis, in photoelectrochemical removal of pollutants, enhancing effect in the degradation process has been often highlighted. Zanoni et al. [17] reported the highest discoloration rate and TOC removal for solution containing Remazol Brilliant Orange 3R at pH 6.0 in presence of 0.5 M of NaCl applying +1.0 V (SCE) to the TiO_2 photoanode. Also, in the case of other dyes or organics, the presence of Cl^- has been found beneficial to accelerate the degradation rate [18,19]. The improvement in the degradation has been explained by the synergistic action of the strong oxidizing species HO•, chlorine-based radicals Cl• and $Cl_2\bullet^-$, and active chlorine species like HClO and Cl_2 that can give a bulk contribution [20,21]. Moreover, at the anode the adsorption of negative charged ions, such as chloride, can be enhanced both by the polarization and the promotion of reactions that can generate local acidic pH variation near the anodic surface.

In this framework, our work is devoted to study the photoelectrochemical degradation of a persistent organic pollutant at two levels of applied potentials and in the presence of different concentrations of chloride under simulated solar light conditions, using TiO_2 nanotubular electrodes. The pollutant selected for the study is the diethylphthalate (DEP). Phthalate esters (PAEs) are a group of widely used plasticizers that can lead to endocrine system disorders, affecting reproductive function, and inducing some tumors [22–24]. Due to their wide utilization and the difficulty to completely remove them with conventional treatment processes, PAEs are ubiquitous persistent organic pollutants in the environment, being the short chain phthalate as DEP, the most detected in surface marine waters, freshwaters, and sediments [25–27]. To the best of our knowledge, only few papers reported on the photoelectrochemical degradation of the diethyl phthalate [28,29]. Moreover, the influence of the presence of chloride during their treatment and the effect of the applied potential are not yet presented by the literature.

2. Materials and Methods

2.1. Preparation of TiO_2 Nanotubes

TiO_2 nanotube electrode (TiO_2-NT) used for the photoelectrochemical degradation of DEP was prepared by electrochemical anodization as reported in our previous work [30]. Briefly, Ti foils (0.25 mm thickness, 99.7% metal basis, Aldrich, St. Louis, MO, USA) were cut in circular disks of 5 cm diameter. After ultrasonic treatments in acetone, isopropanol and methanol (10 min each), Ti was rinsed with deionized water, and dried with a nitrogen stream. The anodization was performed in a two-electrode cylindrical cell made by Teflon (inner dimension: diameter = 4.4 cm and height = 5 cm). The working electrode was located at the bottom of the cell where the electrical contact was an aluminum disc. The exposed geometrical area of the Ti electrodes was 15 cm^2. A platinum titanium grid placed in front of the anode at 1 cm distance constituted the counter electrode.

The anodization was performed in (10%) deionized water/(90%) glycerol solution with 0.14 M of NH_4F at room temperature. A potential ramp was imposed from open circuit voltage (OCV) to 20 V with a scan rate of 100 mVs^{-1}; then the applied potential was maintained at this fixed value for 4 h. TiO_2-NT was annealed in air atmosphere at 400 °C for 1 h to transform the amorphous structure into crystalline one. The phase transformation depends on both the structure morphology and annealing temperature: it has been shown that the anatase-to-rutile transformation starts near 430 °C for the 500 nm long nanotubes [31], while the same transformation has been reported to occur at 550 °C for nanotubes up to 200 nm [32]. In our case, after 1 h at 400 °C, a unique anatase phase was present [33]. The morphological characterization of TiO_2-NT was presented in [30]: the average diameter of tubes ranged between 40–50 nm, while the tube length of around 700 nm was measured.

2.2. Photoelectrochemical Tests

Photoelectrochemical tests were performed in a three-electrode beaker cell using TiO_2 nanotubes as photoanode, a platinized titanium grid as cathode, and a saturated calomel electrode (SCE) as reference. The cell was filled with 100 mL of solution and connected with a potentiostat-galvanostat (Metrhom Autolab 302N, Metrohm, Herisau, Switzerland) controlled by Nova software. The photoanode was irradiated by UV-vis light using a 300 W xenon lamp equipped with air mass (AM) 0 and 1.5 D filters to simulate the solar irradiation.

Photocurrent measurements were carried out by linear sweep voltammetric (LSV) runs, starting from the OCV to 2.5 V at a scan rate of 10 mVs^{-1}, with hand-chopped light. The photocurrent-time measurements were recorded applying a constant potential in the dark for 10 min; afterward, the electrode was exposed to light for 200 s, followed by dark condition.

Photoelectrochemical oxidation of diethyl phthalate was performed under potentiostatic conditions at 0.2 and 1.8 V vs. SCE. The initial concentration of the organic compound was 40 $mg\,dm^{-3}$ and 0.1 M $NaClO_4$ was used as supporting electrolyte. Moreover, different amount of NaCl (1, 100, 500 mM) were added to the solution, to investigate on the effect of chloride concentration during the photoelectrochemical oxidation of DEP. The pH of the solution was neutral. During degradation experiments, samples of electrolyte were withdrawn for qualitative and quantitative analyses of the model organic compound.

2.3. Analytical Methods

Analyses of the model organic compound were carried out by HPLC (Waters), equipped with a column Varian C18 and a dual band UV detector set to 283 and 229 nm. The mobile phase was Acetonitrile and aqueous solution 0.1% H_3PO_4 = 40:60 with a flow rate of 1 $mL\,min^{-1}$.

The oxidant concentration, expressed as µM of active chlorine, was measured using the N,N-diethyl-p-phenylenediamine (DPD) colorimetric method. DPD oxidizes to form a red-violet product, the concentration of which is determined measuring the absorbance at 515 nm.

The trend of mineralization was monitored by measuring the total organic carbon (TOC) by a Shimatzu TOC 500L instrument.

For each sample a repeatability within ±5% has been evaluated.

3. Results and Discussion

Figure 1 shows the trend of polarization curve performed at the TiO_2-NT electrode during LSV in aqueous solution of DEP under irradiation and in the dark.

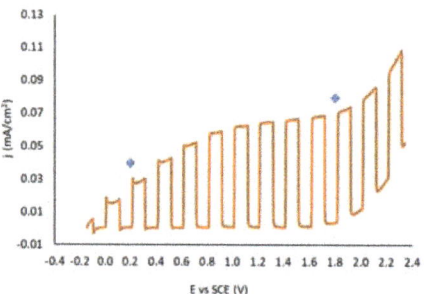

Figure 1. LSV of TiO$_2$-NT performed at 10 mV s^{-1} of scan rate, under dark and irradiation condition. Blue symbols indicate the potentials selected for the degradation runs.

A typical trend is observed, with an onset potential of −0.25 V, followed by an ohmic behavior of the system, in which the positive influence of the potential is strictly connected to the increase in the space charge depletion region of the semiconductor; in the central range of potential (0.7–1.8 V) the saturation of the current is reached, in which increase of the potential is no more effective in terms of a corresponding increasing of the current. In the final range, at potentials higher than the value of band gap of the semiconductor, the barrier breakdown effect could be responsible for the sharp rising in the photocurrent along with the dark current contribution [30].

The degradation tests have been performed selecting two applied potentials: the first one in the ohmic region and the second one in the saturation region. The two blue diamonds in Figure 1 indicate the values of potential selected.

Figure 2a shows the trend with time of the DEP concentration, normalized with respect to the initial concentration, during electrolysis at the two different potentials. For comparison, the trend with time of the DEP concentration at the open circuit potential in the dark was also reported in the same figure: no significant adsorption of DEP on the electrode surface was detected that can be explained considering the neutral pH of the solution, the iso-electrical point of TiO$_2$ located around pH = 6, and the non-ionic nature of the molecule of DEP. When the runs were performed in potentiostatic conditions and under illumination, the concentration of DEP decreased, being the highest reaction rate achieved at 1.8 V.

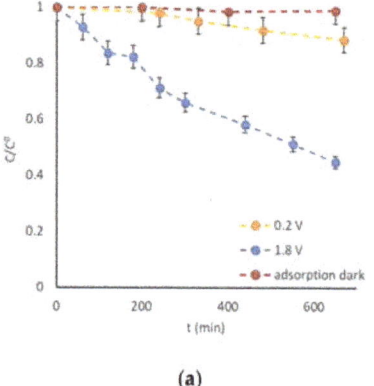

(a)

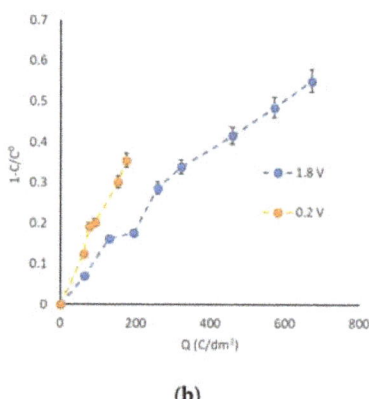

(b)

Figure 2. (a) Trends with time of the concentration of DEP, normalized to the initial concentration C^0, during runs performed with solutions containing 40 mg dm^{-3} DEP in 0.1 M NaClO$_4$ as supporting electrolyte at different applied potentials. (b) Fraction of reactant removed as a function of the specific charge supplied during the related runs.

However, since the mean current intensity measured during the potentiostatic runs was 0.1 mA at 0.2 V and 1.2 mA at 1.8 V, it could be useful to compare the trend of fraction of the removed reactant as a function of the specific supplied charge (Figure 2b): in this case, the highest yield of the removal process is measured at the lowest potential, indicating that most of the charge passed at 1.8 V has been used for the side reaction of water oxidation.

An analogous behavior was observed in our previous work, where the photo-electrocatalytic degradation of 2,4-dichlorophenoxyacetic acid was investigated: higher efficiency and slower kinetics of degradation were detected in the ohmic region of the polarization curve with respect to those in the saturation region [30].

Degradation curves of DEP at various chloride concentration at the two applied potentials are shown in Figure 3a,b as semilogarithmic plots. A linear trend of $\ln(C/C^0)$ vs. time is observed under all the experimental conditions, indicating that a pseudo-first order kinetics could be used to interpret the data, as follows:

$$dC/dt = -k_{app} C \tag{1}$$

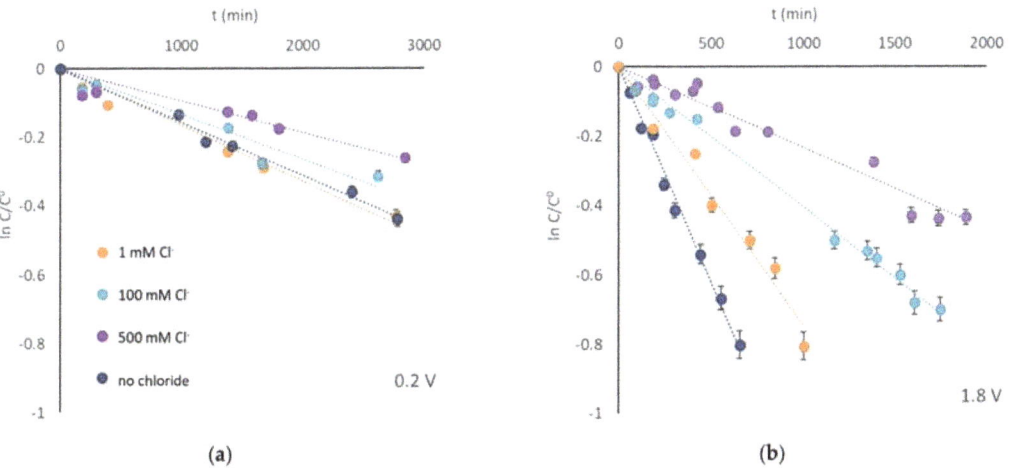

Figure 3. Trends with time of $\ln C/C^0$ during photoelectrochemical degradations using solutions containing 40 mg dm^{-3} DEP, 0.1 M NaClO$_4$, and different chloride concentration. (**a**) Applied potential: 0.2 V; (**b**) applied potential: 1.8 V.

The values of the apparent kinetic constant k_{app}, evaluated from the slope of each straight line at the relevant operative conditions are reported in Figure 4, as a function of the chloride concentration. As already observed in absence of chloride, the fastest kinetics of the reactant removal are obtained at 1.8 V for each level of chloride concentration. Moreover, at 0.2 V, the increase of chloride concentration scarcely affects the reaction rate, except for 500 mM of Cl$^-$, which halves the k_{app}. At 1.8 V, the effect of chloride is more evident: at 1 mM of Cl$^-$ the k_{app} is reduced by 40% while at 500 mM of Cl$^-$ by 80%, in respect to the k_{app} evaluated without chloride.

Figure 5 shows the trend of the ratio between k_{app} evaluated at 1.8 V and that at 0.2 V measured at different chloride concentrations. In absence of chloride, an increment of one order of magnitude is obtained, while in presence of the highest concentration of chloride k_{app} increases of two-fold when the potential values change from 0.2 to 1.8 V. This behavior indicates that the higher the potential, the higher is the negative effect of the concentration of chloride.

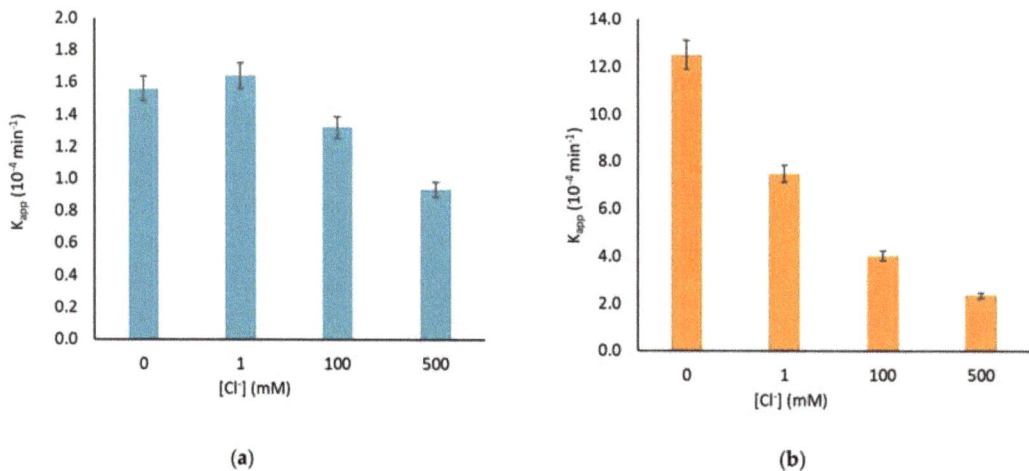

Figure 4. Pseudo-first order kinetic constants of the reactant removal process performed in solutions of 40 mg dm^{-3} of DEP, 0.1 M NaClO$_4$, and different chloride concentration. (**a**) Applied potential: 0.2 V; (**b**) applied potential: 1.8 V.

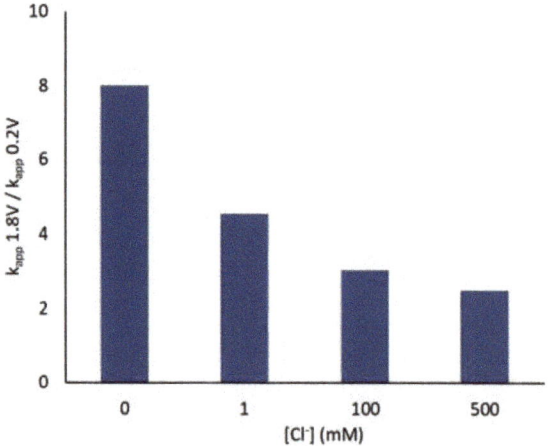

Figure 5. Ratio between the apparent kinetic constant evaluated at 1.8 and 0.2 V for different chloride concentrations.

The inhibiting effect observed in presence of Cl$^-$ agrees with observations reported for photocatalytic processes at TiO$_2$-based materials. Several mechanisms have been proposed to explain the inhibiting effect on the photocatalytic degradation [13]:

(1) scavenging of holes or HO• radicals by chloride ions [34–36].
(2) blocking of active surface sites by chloride ions [3,20,37].
(3) chloride acting as surface-charge-recombination center for photogenerated charge carriers [38].

Moreover, due to the complexity of the processes, a combination of mechanisms is often claimed to explain the inhibiting effect [5,13,39–41].

In the case of a photo-electrochemical process, also the effect of the applied potential should be considered, as well as the pH modification due to the side reactions that occur to a greater or lesser extent depending on the applied potential.

In order to verify the effect of the concentration of chloride and the applied potential on the behavior of the semiconductor, photocurrent transients have been recorded applying different potential and varying the chloride concentration during chopped light chronoamperometries.

Figure 6 shows the results obtained without chloride. For TiO_2 nanotubes, the thickness of the wall can be determinant for the extension of the space charge depletion layer; this in turn, can be relevant for the recombination phenomena, which are strictly connected to the applied potential. As can be seen, at the lowest potential, a typical spike of the anodic current is observed, followed by an exponential decrease of the photocurrent with time until a stationary value is reached. The positive spike is no more visible at the highest potential. According to the literature [42], the positive current transient when the light is turned on represents the accumulation of holes at the electrode/electrolyte interface without injection to the electrolyte. Since any fast faradic reaction is occurring, the charge recombination is responsible for the subsequent decrease of the measured current.

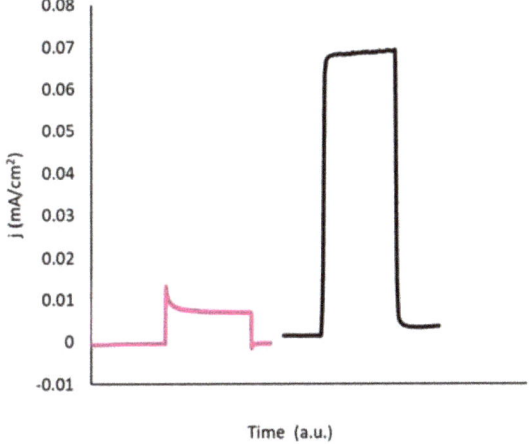

Figure 6. Potentiostatic tests performed with solution containing 0.1 M $NaClO_4$ at 0.2 (**pink**) and 1.8 V (**black**).

At low potentials, when we operate in the ohmic region of the polarization curve, where the charge depletion layer thickness is not fully developed inside the nanotubes wall, the photogenerated holes may rapidly recombine in the regions of the material that do not experience beneficial space charge effects, i.e., that are non-depleted of the majority carriers (electrons). When the experiment is performed at the highest potential (in the saturation region of the polarization curve) the depletion layer extends in the whole wall of nanotubes and the recombination is suppressed.

Photocurrent transients in presence of chloride are reported in Figure 7.

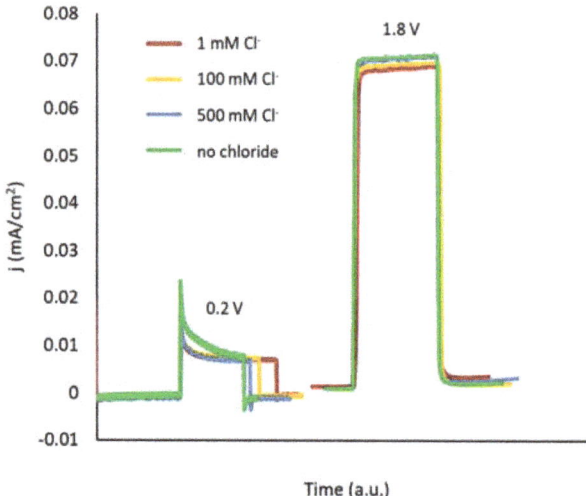

Figure 7. Potentiostatic tests performed with solution containing 0.1 M NaClO$_4$ and different concentration of chloride ions at 0.2 and 1.8 V.

At low applied potential, the rate of the photocurrent decreasing (i.e., the rate of the charge recombination process) is scarcely influenced by the presence of chloride, when they are present at low concentration levels: overlapped curves are obtained related to the runs performed at 0, 1, and 100 mM of Cl$^-$ ions. Only at 500 mM of Cl$^-$, slower decay of the photocurrent can be observed, indicating an inhibition of the recombination processes. Moreover, when the light was turned off, negative current transients were observed for high chloride concentration. Negative spikes were often detected during photocurrent transient of semiconductors and can be related to slower electron/hole pairs recombination due to the presence of holes trapped in the surface [43].

These transients in photocurrent can be explained as follows: at lower chloride concentration, the charge recombination prevails since chloride is poorly adsorbed onto the semiconductor electrode, so it is not able to react with the photogenerated holes faster than the electrons. However, at the highest concentration of chloride, it is likely that the adsorption effect would predominate, so that chloride can act as hole scavenger, according to the following adsorption phenomena:

$$TiO_2\text{-}h^+ + Cl^- = TiO_2\text{-}Cl_{ads} \qquad (2)$$

This process promotes the separation of electron-hole pair limiting the charge recombination as suggested by other authors [20,44,45].

At the highest potential, the recombination is suppressed, and positive transient and negative spikes disappear also in presence of high chloride ions. Moreover, very small increment in the steady state photocurrent was observed, increasing the concentration of chloride. So, at 0.2 V, the highest variation in the value of k_{app} obtained at 500 mM of chloride, can be connected to the blocking effect of adsorbed Cl$^-$ and the competitive adsorption, with respect to water molecules, which reduces the formation of HO• radicals.

Similar considerations should be done also to explain the result at 1.8 V, but, as we noticed, the inhibiting effect at this potential is evident also at low concentration of chloride. This can be explained by considering two aspects connected to the applied potential: the electrode works in a region of potential where the oxygen evolution reaction occurs to a large extent, so that a local acidic pH near the surface can generate a positive charge (pH < isoelectric point). Moreover, the application of high anodic potentials can generate a build-up of a positive surface charge. In this condition, the competitive adsorption or blocking of active surface sites by chloride anions will be favored due to the electrostatic attraction of Cl^-, also at low concentration of chloride.

The adsorbed chloride can react to form chlorine by the following reaction [17,44]:

$$TiO_2\text{-}Cl_{ads} + Cl^- \rightarrow Cl_2 + TiO_2 + e^- \tag{3}$$

Dissolved chlorine reacts with water to give hypochlorous acid and hypochlorite ions (Equations (3) and (4)), being the distribution of the three forms of active chlorine dependent on pH:

$$Cl_2 + H_2O \rightarrow Cl^- + HClO + H^+ \tag{4}$$

$$HClO \leftrightarrow = H^+ + ClO^- \tag{5}$$

Chlorine-based oxidants (active chlorine) have been detected during the photo-electrochemical degradation of DEP in different operating conditions.

At 0.2 V after 130 C dm^{-3} of supplied charge, 2.0 and 4.2 µM of active chlorine concentrations were detected at 100 and 500 mM of Cl^-, respectively. These small amounts agree with the poor adsorption of chloride at this value of applied potential. At 1.8 V, higher concentration of active chlorine was detected. As an example, the trend with time of the concentration of active chlorine obtained during DEP degradation in presence of 100 mM of Cl^- is reported in Figure 8. The higher amount of active chlorine confirms a better reactivity of chloride with the positively charged surface of TiO_2 at 1.8 V.

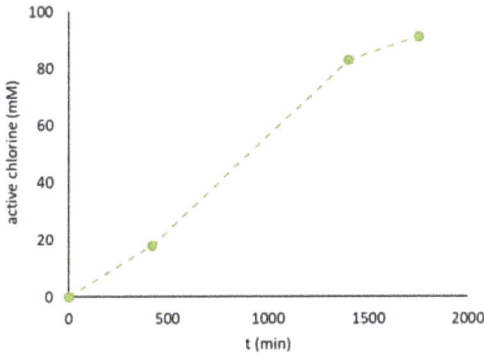

Figure 8. Trends with time of the concentration of active chlorine produced during a degradation run at 1.8 V with solution containing 40 mg dm^{-3} of DEP, 0.1 M $NaClO_4$, and 100 mM of Cl^-.

The formation of chlorine-based oxidants during photoelectrochemical treatment of water containing chloride has been studied by several authors: some of them found that the presence of Cl^- suppressed the degradation rate of organic pollutants, while others found opposed result [44].

The positive effect was generally observed when the active chlorine was able to give a bulk contribution to the reaction, i.e., in the cases where the organic pollutants can be oxidized also by active chlorine. For example, during photo-electrochemical discoloration of solutions containing Methylene Blue, low pH, and high concentration of Cl^- were beneficial [18]. Also, Zanoni et al. [17] found the highest TOC removal for solution containing Remazol Brilliant Orange 3R, working at pH 6.0, 1.0 M NaCl, when the photoelectrode was biased at +1 V (versus SCE).

In our case, the formation of active chlorine seems not sufficient to contribute to the overall reaction rate at such an extent to make up for the negative effect.

Some specific tests were performed to evaluate the effectiveness of the photo-electrogenerated active chlorine on the DEP degradation. To this aim, during photoelectrocatalytic degradation runs, the light was turned off and the application of bias potential was stopped. In this condition, in the solution, 25 µM of active chlorine accumulated, and residual 26 mg dm^{-3} of DEP were present: the solution was monitored by following the concentration of the residual DEP with time. Negligible variation in the concentration of DEP was found after two hours indicating that the HO• radicals may be considered as the main factor responsible for the degradation, while active chlorine seems to give a not significant contribution to the overall oxidation rate. Similar behavior was found during the electrochemical degradation of the dimethyl phthalate ester on a fluoride-doped Ti/β-PbO$_2$ anode: the lower removal of the pollutant in the presence of chloride ions was explained considering the lower reactivity of dimethyl phthalate with chlorine radical species in respect to hydroxyl radicals. Also, the active chlorine can react with HO• radicals thus reducing their availability for organic oxidation [46].

The low reactivity of active chlorine towards DEP obtained in our experimental conditions may indicate that the formation of harmful chlorinated intermediates is unlikely, even if the possible reaction of DEP intermediates with active chlorine during the runs cannot be excluded. Table 1 reports the ratio (φ) between the removal percentages of TOC and DEP evaluated at the end of each run, which indicates the level of total mineralization as defined by the following equation [47]:

$$\varphi = \frac{\%[TOC]_{removal}}{\%[DEP]_{removal}} \quad (6)$$

Table 1. TOC removal and φ evaluated at the end of each run.

Applied Potential (V)	[Cl$^-$] mM	TOC Removal	φ
0.2 V	0	31%	0.87
	1	28%	0.87
	100	26%	0.96
	500	11%	0.98
1.8 V	0	53%	0.96
	1	46%	0.99
	100	49%	1.00
	500	39%	1.00

At 1.8 V, a higher degree of mineralization was evaluated at the end of the runs, in which φ approached the unity. However, at 0.2 V, the high values of φ also indicate that the possible intermediates are almost completely removed.

4. Conclusions

In this work, the photoelectrochemical degradation of diethyl phthalate has been studied at two levels of applied potentials and in the presence of different concentrations of chloride under simulated solar light conditions, using TiO$_2$ nanotubular electrodes. The process is able to remove DEP following a pseudo-first order kinetics: values of k$_{app}$ of 1.25×10^{-3} min^{-1} and 1.56×10^{-4} min^{-1} were obtained at applied potentials of 1.8 and 0.2 V, respectively. Higher current efficiency and slower kinetics of degradation were detected in the ohmic region of the polarization curve at 0.2 V. The presence of chloride ions in the solution affects the degradation rate to different extents depending of the applied potential: the higher the potential, the higher the negative effect of the increase of chloride concentration. The presence of 500 mM of Cl$^-$ halves the k$_{app}$ at 0.2 V, while at 1.8 V its value decreases to 1.56×10^{-4} min^{-1}. This behavior can be connected to the blocking effect of adsorbed Cl$^-$ and the competitive adsorption, with respect to water molecules, which reduces the formation of HO• radicals: at 0.2 V, the adsorption of chloride predominates only at the highest concentration of chloride, while at 1.8 V, the positive surface charge due to the applied potential and the possible acidification of the anodic layer allow the adsorption also at low chloride concentrations.

Author Contributions: Conceptualization, A.V. and S.P.; methodology, M.M.; validation, L.M., A.V. and S.P.; formal analysis, L.M.; investigation, L.M.; writing—original draft preparation, A.V. and S.P.; writing—review and editing, S.P., L.M. and M.M. All authors have read and agreed to the published version of the manuscript.

Funding: This paper is part of the research project funded by P.O.R. SARDEGNA F.S.E. 2014–2020—Axis III Education and Training, Thematic Goal 10, Specific goal 10.5, Action partnership agreement 10.5.12—"Call for funding of research projects—Year 2017".

Data Availability Statement: Data is contained within the article.

Conflicts of Interest: The authors declare no conflict of interest.

References

1. Palmas, S.; Mais, L.; Mascia, M.; Vacca, A. Trend in using TiO_2 nanotubes as photoelectrodes in PEC processes for wastewater treatment. *Curr. Opin. Electrochem.* **2021**, *28*, 100699.
2. van de Krol, R. Principles of Photoelectrochemical Cells. In *Photoelectrochemical Hydrogen Production*; van de Krol, R., Grätzel, M., Eds.; Electronic Materials: Science & Technology; Springer: Boston, MA, USA, 2012; Volume 102, pp. 13–67.
3. Krivec, M.; Dillert, R.; Bahnemann, D.W.; Mehle, A.; Strancar, J.; Drazic, G. The nature of chlorine-inhibition of photocatalytic degradation of dichloroacetic acid in a TiO_2-based microreactor. *Phys. Chem. Chem. Phys.* **2014**, *16*, 14867. [CrossRef] [PubMed]
4. Lin, L.; Jiang, W.; Chen, L.; Xu, P.; Wang, H. Treatment of produced water with photocatalysis: Recent advances, affecting factors and future research prospects. *Catalysts* **2020**, *10*, 924. [CrossRef]
5. Piscopo, A.; Robert, D.; Weber, J.V. Influence of pH and chloride anion on the photocatalytic degradation of organic compounds Part I. Effect on the benzamide and para-hydroxybenzoic acid in TiO_2 aqueous solution. *Appl. Catal. B Environ.* **2001**, *35*, 117–124. [CrossRef]
6. Selcuk, H.; Bekbolet, M. Photocatalytic and photoelectrocatalytic humic acid removal and selectivity of TiO_2 coated photoanode. *Chemosphere* **2008**, *73*, 854–858. [CrossRef]
7. Chong, M.N.; Jin, B.; Chow, C.W.K.; Saint, C. Recent developments in photocatalytic water treatment technology: A review. *Water Res.* **2010**, *44*, 2997–3027. [CrossRef]
8. Zhou, X.; Zheng, Y.; Zhou, J.; Zhou, S. Degradation kinetics of photoelectrocatalysis on landfill leachate using codoped TiO_2/Ti photoelectrodes. *J. Nanomater.* **2015**, *2015*, 1–11. [CrossRef]
9. Konstantinou, I.K.; Albanis, T.A. TiO_2-assisted photocatalytic degradation of azo dyes in aqueous solution: Kinetic and mechanistic investigations: A review. *Appl. Catal. B Environ.* **2004**, *49*, 1–14. [CrossRef]
10. Moser, J.; Gratzel, M. Photoelectrochemistry with colloidal semiconductors; laser studies of halide oxidation in colloidal dispersions of TiO_2 and α-Fe_2O_3. *Helv. Chim. Acta* **1982**, *65*, 1436–1444. [CrossRef]
11. Ahmad, R.; Ahmad, Z.; Khan, A.U.; Mastoi, N.R.; Aslam, M.; Kim, J. Photocatalytic systems as an advanced environmental remediation: Recent developments, limitations and new avenues for applications. *J. Environ. Chem. Eng.* **2016**, *4*, 4143–4164. [CrossRef]
12. Sirtori, C.; Agüera, A.; Gernjak, W.; Malato, S. Effect of water-matrix composition on Trimethoprim solar photodegradation kinetics and pathways. *Water Res.* **2010**, *44*, 2735–2744. [CrossRef]
13. Brüninghoff, R.; van Duijne, A.K.; Braakhuis, L.; Saha, P.; Jeremiasse, A.W.; Mei, T.; Mul, G. Comparative Analysis of Photocatalytic and Electrochemical Degradation of 4-Ethylphenol in Saline Conditions. *Environ. Sci. Technol.* **2019**, *53*, 8725–8735. [CrossRef]
14. Peng, Y.P.; Yassitepe, E.; Yeh, Y.T.; Ruzybayev, I.; Shah, S.I.; Huang, C.P. Photoelectrochemical degradation of azo dye over pulsed laser deposited nitrogen-doped TiO_2 thin film. *Appl. Catal. B Environ.* **2012**, *125*, 465–472. [CrossRef]
15. Vacca, A.; Mais, L.; Mascia, M.; Usai, E.M.; Palmas, S. Design of experiment for the optimization of pesticide removal from wastewater by photo-electrochemical oxidation with TiO_2 nanotubes. *Catalysts* **2020**, *10*, 512. [CrossRef]
16. Pan, X.; Yang, M.Q.; Fu, X.; Zhang, N.; Xu, Y.J. Defective TiO_2 with oxygen vacancies: Synthesis, properties and photocatalytic applications. *Nanoscale* **2013**, *5*, 3601–3614. [CrossRef] [PubMed]
17. Zanoni, M.V.B.; Sene, J.J.; Anderson, M.A. Photoelectrocatalytic degradation of Remazol Brilliant Orange 3R on titanium dioxide thin-film electrode. *J. Photochem. Photobiol. A Chem.* **2003**, *157*, 55–63. [CrossRef]
18. Wu, X.; Huang, Z.; Liu, Y.; Fang, M. Investigation on the Photoelectrocatalytic Activity of Well-Aligned TiO_2 Nanotube Arrays. *Int. J. Photoenergy* **2012**, *2012*, 832516. [CrossRef]
19. An, T.; Zhang, W.; Xiao, X.; Sheng, G.; Fu, J.; Zhu, X. Photoelectrocatalytic degradation of quinoline with a novel three-dimensional electrode-packed bed photocatalytic reactor. *J. Photochem. Photobiol. A Chem.* **2004**, *161*, 233–242. [CrossRef]
20. Zhang, W.; An, T.; Cui, M.; Sheng, G.; Fu, J. Effects of anions on the photocatalytic and photoelectrocatalytic degradation of reactive dye in a packed-bed reactor. *J. Chem. Technol. Biotechnol.* **2005**, *80*, 223–229. [CrossRef]
21. Zanoni, M.V.B.; Sene, J.J.; Selcuk, H.; Anderson, M.A. Photoelectrocatalytic production of active chlorine on nanocrystalline titanium dioxide thin-film electrodes. *Environ. Sci. Technol.* **2004**, *38*, 3203–3208. [CrossRef]

22. Xu, B.; Gao, N.Y.; Sun, X.F.; Xia, S.J.; Rui, M.; Simonnot, M.O.; Causserand, C.; Zhao, J.F. Photochemical degradation of diethyl phthalate with UV/H_2O_2. *J. Hazard. Mater.* **2007**, *139*, 132–139. [CrossRef]
23. Chang, B.V.; Liao, C.S.; Yuan, S.Y. Anaerobic degradation of diethyl phthalate, di-n-butyl phthalate, and di-(2-ethylhexyl) phthalate from river sediment in Taiwan. *Chemosphere* **2005**, *58*, 1067–1601. [CrossRef]
24. Wang, H.; Sun, D.Z.; Bian, Z.Y. Degradation mechanism of diethyl phthalate with electrogenerated hydroxyl radical on a Pd/C gas-diffusion electrode. *J. Hazard. Mater.* **2010**, *180*, 710–715. [CrossRef] [PubMed]
25. Staples, C.A.; Peterson, D.R.; Parkarton, T.F.; Adams, W.J. The environmental fate of phthalate esters: A literature review. *Chemosphere* **1997**, *35*, 667–749. [CrossRef]
26. Staples, C.A.; Parkerton, T.F.; Peterson, D.R. A risk assessment of selected phthalate esters in North American and western European surface waters. *Chemosphere* **2000**, *40*, 885–891. [CrossRef]
27. Wang, R.; Ji, M.; Zhai, H.; Liub, Y. Occurrence of phthalate esters and microplastics in urban secondary effluents, receiving water bodies and reclaimed water. *Sci. Total Environ.* **2020**, *737*, 140219. [CrossRef]
28. Ma, F.; Shi, T.; Gao, J.; Chen, L.; Guo, W.; Guo, Y.; Wang, S. Comparison and understanding of the different simulated sunlight photocatalytic activity between the saturated and monovacant Keggin unit functionalized titania materials. *Colloids Surf. A Physicochem. Eng. Asp.* **2012**, *401*, 116–125. [CrossRef]
29. Cai, J.; Niu, B.; Zhao, H.; Zhao, G. Selective photoelectrocatalytic removal for group-targets of phthalic esters. *Environ. Sci. Technol.* **2021**, *55*, 2618–2627. [CrossRef]
30. Vacca, A.; Mais, L.; Mascia, M.; Usai, M.E.; Rodriguez, J.; Palmas, S. Mechanistic insights into 2,4-D photoelectrocatalytic removal from water with TiO_2 nanotubes under dark and solar light irradiation. *J. Hazard. Mater.* **2021**, *412*, 125202. [CrossRef] [PubMed]
31. Varghese, O.K.; Gong, D.W.; Paulose, M.; Grimes, C.A.; Dickey, E.C. Crystallization and high-temperature structural stability of titanium oxide nanotube arrays. *J. Mater. Res.* **2003**, *18*, 156–165. [CrossRef]
32. Allam, N.K.; El-Sayed, M.A. Photoelectrochemical water oxidation characteristics of Anodically Fabricated TiO_2 nanotube arrays: Structural and optical properties. *J. Phys. Chem. C* **2010**, *114*, 12024–12029. [CrossRef]
33. Palmas, S.; Da Pozzo, A.; Mascia, M.; Vacca, A.; Ardu, A.; Matarrese, R.; Nova, I. Effect of the preparation conditions on the performance of TiO2 nanotube arrays obtained by electrochemical oxidation. *Int. J. Hydrogen Energy* **2011**, *36*, 8894–8901. [CrossRef]
34. Calza, P.; Pelizzetti, E. Photocatalytic transformation of organic compounds in the presence of inorganic ions. *Pure Appl. Chem.* **2001**, *73*, 1839–1848. [CrossRef]
35. Yang, S.; Chen, Y.; Lou, L.; Wu, X. involvement of chloride anion in photocatalytic process. *J. Environ. Sci.* **2005**, *17*, 761–765.
36. Burns, R.A.; Crittenden, J.C.; Hand, D.W.; Selzer, V.H.; Sutter, L.L.; Salman, S.R. Effect of inorganic ions in heterogeneous photocatalysis of TCE. *J. Environ. Eng.* **1999**, *125*, 77–85. [CrossRef]
37. Chen, H.Y.; Zahraa, O.; Bouchy, M. Inhibition of the adsorption and photocatalytic degradation of an organic contaminant in an aqueous suspension of TiO2 by inorganic ions. *J. Photochem. Photobiol. A* **1997**, *108*, 37–44. [CrossRef]
38. Sunada, F.; Heller, A. Effects of water, salt water, and silicone overcoating on the TiO_2 photocatalyst on the rates and products of photocatalytic oxidation of liquid 3-octanol and 3-octanone. *Environ. Sci. Technol.* **1998**, *32*, 282–286. [CrossRef]
39. Abdullah, M.; Low, G.K.C.; Matthews, R.W. Effects of common inorganic anions on rates of photocatalytic oxidation of organic carbon over illuminated titanium dioxide. *J. Phys. Chem.* **1990**, *94*, 6820–6825. [CrossRef]
40. Azevedo, E.B.; de Aquino Neto, F.R.; Dezotti, M. TiO_2-photocatalyzed degradation of phenol in saline media: Lumped kinetics, intermediates, and acute toxicity. *Appl. Catal. B* **2004**, *54*, 165–173. [CrossRef]
41. Adishkumar, S.; Kanmani, S.; Rajesh Banu, J. Solar photocatalytic treatment of phenolic wastewaters: Influence of chlorides, sulphates, aeration, liquid volume and solar light intensity. *Desalin. Water Treat.* **2014**, *52*, 7957–7963. [CrossRef]
42. Denisov, N.; Yoo, J.E.; Schmuki, P. Effect of different hole scavengers on the photoelectrochemical properties and photocatalytic hydrogen evolution performance of pristine and Pt-decorated TiO_2 nanotubes. *Electrochim. Acta* **2019**, *319*, 61–71. [CrossRef]
43. Peter, L.M.; Upul Wijayantha, K.G.; Tahir, A.A. Kinetics of light-driven oxygen evolution at α-Fe_2O_3 electrodes. *Faraday Discuss.* **2012**, *155*, 309–322. [CrossRef]
44. Xiao, S.; Qu, J.; Liu, H.; Zhao, X.; Wan, D. Fabrication of TiO_2/Ti nanotube electrode and the photoelectrochemical behaviors in NaCl solutions. *J. Solid State Electrochem.* **2009**, *13*, 1959–1964. [CrossRef]
45. Spadavecchia, F.; Ardizzone, S.; Cappelletti, G.; Falciola, L.; Ceotto, M.; Lotti, D. Investigation and optimization of photocurrent transient measurements on nano-TiO_2. *J. Appl. Electrochem.* **2013**, *43*, 217–225. [CrossRef]
46. Souza, F.L.; Aquino, J.M.; Irikura, K.; Miwa, D.W.; Rodrigo, M.A.; Motheo, A.J. Electrochemical degradation of the dimethyl phthalate ester on a fluoride-doped Ti/b-PbO_2 anode. *Chemosphere* **2014**, *109*, 187–194. [CrossRef] [PubMed]
47. Miwa, D.W.; Malpass, G.R.P.; Machado, S.A.S.; Motheo, A.J. Electrochemical degradation of carbaryl on oxide electrodes. *Water Res.* **2006**, *40*, 3281–3289. [CrossRef] [PubMed]

Article

Analysis of Photocatalytic Degradation of Phenol with Exfoliated Graphitic Carbon Nitride and Light-Emitting Diodes Using Response Surface Methodology

Adeem Ghaffar Rana [1,2] and Mirjana Minceva [1,*]

[1] Biothermodynamics, TUM School of Life Sciences, Technical University of Munich, Maximus-von-Imhof-Forum 2, 85354 Freising, Germany; adeem.rana@tum.de

[2] Department of Chemical, Polymer, and Composite Materials Engineering, University of Engineering and Technology (UET), Lahore 39161, Pakistan

* Correspondence: mirjana.minceva@tum.de; Tel.: +49-8161716170

Abstract: Response surface methodology (RSM) involving a Box–Benkhen design (BBD) was employed to analyze the photocatalytic degradation of phenol using exfoliated graphitic carbon nitride (g-C_3N_4) and light-emitting diodes (wavelength = 430 nm). The interaction between three parameters, namely, catalyst concentration (0.25–0.75 g/L), pollutant concentration (20–100 ppm), and pH of the solution (3–10), was examined and modeled. An empirical regression quadratic model was developed to relate the phenol degradation efficiency with these three parameters. Analysis of variance (ANOVA) was then applied to examine the significance of the model; this showed that the model is significant with an insignificant lack of fit and an R^2 of 0.96. The statistical analysis demonstrated that, in the studied range, phenol concentration considerably affected phenol degradation. The RSM model shows a significant correlation between predicted and experimental values of photocatalytic degradation of phenol. The model's accuracy was tested for 50 ppm of phenol under optimal conditions involving a catalyst concentration of 0.4 g/L catalysts and a solution pH of 6.5. The model predicted a degradation efficiency of 88.62%, whereas the experimentally achieved efficiency was 83.75%.

Keywords: g-C_3N_4; photocatalysis; response surface methodology; wastewater treatment; phenol

1. Introduction

For all living beings, water is considered to be the most important resource. Easy access to clean water is one of the biggest challenges for mankind. In the last few decades, advancements in science, technology, and industrialization have led to considerable benefits to mankind but at the cost of a more polluted environment, particularly water [1]. There are multiple categories of pollutants in water, such as heavy metals, dyes, pesticides, pharmaceuticals, and other organic pollutants. Amongst organic pollutants, phenolic compounds, with ~3 million tons of global production, are an emerging contaminant detected in water [1–4].

Phenols or phenolics are essential because of their wide range of applications in the processing and manufacturing industry. However, the ecosystem's contamination by phenolics is concerning because of the adverse implications on human health such as their endocrine-disrupting abilities and carcinogenic behavior [1,5,6]. Moreover, these chemicals cause environmental issues such as water hardness, pH change, and a decrease in dissolved oxygen level. Furthermore, the Environmental Protection Agency (EPA) and the European Union (EU) have included a few phenols in their priority pollutants list. It is necessary to make this polluted water containing phenols and other pollutants suitable for human use and aquatic life using certain techniques to minimize the usage of these chemicals [5].

The removal of phenolic compounds from wastewater has attracted considerable attention from researchers [5]. Many biological, chemical, and physical techniques such as membrane filtration, coagulation–flocculation, adsorption [7,8], ion exchange, bacterial and fungal biosorption [9], aerobic and anaerobic processes [10] are used for phenol removal. In these processes, there are many constraints such as high cost, and low efficiency; furthermore, these methods do not completely remove phenol from wastewater [11,12]. Moreover, using these techniques, phenol is transferred from wastewater to a solid phase that requires treatment for safe disposal, which leads to additional cost for the whole process. Thus, it is necessary to develop an alternative effective and cost-efficient method for phenol removal from wastewater.

Advanced oxidative processes (AOP) are successful for achieving the complete removal of pollutants [13]. The degradation process using AOP can be performed in several ways, such as using only oxidizing agents, light irradiance in addition with oxidizing agents, and photocatalysis [14]. For all these processes, the degradation process is conducted using OH^- radicals that are generated during the oxidation reaction. Among these processes, photocatalysis has attracted considerable interest because it can harvest solar light with the help of semiconductor materials (catalysts). The catalysts can help solve environmental issues related to water contaminations; these semiconductor materials are nontoxic and efficient. Note that different semiconductor materials such as ZnO [15], TiO_2 [16], SiO_2, Al_2O_3 [8], and $g-C_3N_4$ [17,18], are used for environmental applications in photocatalysis; these have considerable advantages because of the large surface areas, adsorption capacities, and better absorption of light. Among these materials, $g-C_3N_4$ offers improved visible light absorption [17,19–21].

$g-C_3N_4$, a polymeric semiconductor, composed of C, N, and H, has gained considerable interest from researchers for novel generation of photocatalysts because of its widespread catalytic uses in oxidation and reduction processes, such as pollutant degradation, water splitting, and CO_2 reduction. These materials have been extensively used for environmental remediation because they are easy to synthesize, metal-free, inexpensive, and easily available [22–24]. Furthermore, $g-C_3N_4$ possesses higher thermal and chemical stability because of π-conjugated frameworks connecting the 2D layered structure of tri-s-triazine building blocks. $g-C_3N_4$ can be activated by visible light of 420–460 nm because of its low bandgap energy (2.7 eV) [25,26]. There are, however, certain challenges associated with the application of $g-C_3N_4$ in phenol removal such as low surface area, fast recombination rate, and low conductivity, thus resulting in lower efficiency. To overcome these limitations, multiple strategies have been used to improve the surface electronic structures and activity of the bulk $g-C_3N_4$ in visible light. To improve the activity of pristine $g-C_3N_4$, strategies such as metal and non-metal doping, exfoliation, hard and soft templating, and metal oxide heterojunctions have been used [27–31].

Factors affecting the removal efficiency can be tuned by the morphology and/or chemistry of the catalyst and by optimizing the operating parameters. Multiple operating parameters play an important role in the photocatalytic degradation process, thus making their optimization important for achieving good photocatalytic degradation of the target pollutant. Response surface methodology (RSM) is one of the most commonly applied optimization techniques; it is a powerful optimization tool for an experimental design that efficiently helps in systemic analysis [5,11,14]. RSM uses mathematics and statistics to analyze the relative significance of influencing factors on the response of the studied system. RSM is suitable for predicting the effect of individual experimental operating parameters, in addition to locating interactions between parameters and their impact on a response variable. RSM uses a systematic technique to simultaneously vary all parameters and evaluate the influence of these parameters on photocatalytic degradation [32,33]. The greatest advantage of RSM lies in the systematic approach for the experimental design, which mostly requires fewer experiments, thus reducing the time required and thereby being more economical. For designing these experiments, a central composite design (CCD) [3] and Box–Benkhen design (BBD) [11,12] are most commonly used. For the same number of parameters, BBD requires fewer experiments than CCD [3]; therefore, in this study, BBD is selected as a preferred design approach.

The objective of this study was to analyze the photocatalytic degradation of phenol with metal-free g-C_3N_4 and visible LED light and to model the process using RSM. In this study, the operating parameters considered were catalyst concentration, phenol concentration, and pH of the solution. BBD was used for the experimental design and RSM was applied to determine the mathematical relationship between operating parameters and phenol degradation. Finally, the correlation determined by RSM was experimentally validated.

2. Materials and Methods

2.1. Chemicals and Materials

Melamine ($C_3H_6N_6$, 99%) was purchased from Alfa Aesar. Phenol (C_6H_5OH, 99%) was purchased from Merck. Acetonitrile (C_2H_3N, 99.99%) and ultra-pure water for high-performance liquid chromatography (HPLC) were purchased from Sigma Aldrich. NaOH and HCl were purchased from VWR chemicals. All chemicals used were of analytical grade and used as-received without any further purification.

2.2. Photocatalyst Synthesis

Photocatalyst was prepared as per the procedure used in our previous study [18]; the synthesis process is briefly reported here. Melamine was placed in a muffle furnace (Carbolite Gero, GPC 1200, Derbyshire, UK) in a closed crucible to prepare bulk g-C_3N_4 using thermal decomposition. The synthesis process comprised two steps: A heating ramp rate of 2 °C min^{-1} was programmed up to 450 °C; this temperature was maintained for 2 h. Then, the temperature was increased to 550 °C using a heating ramp rate of 2 °C min^{-1} and then maintained for 4 h. The material synthesized was crushed in mortar after cooling, then rinsed with ultrapure water, and dried overnight at 80 °C. The exfoliation process was conducted in an open crucible at 500 °C for 2 h at a heating ramp rate of 2 °C min^{-1} in a muffle furnace.

2.3. Characterization of the Photocatalyst

Fourier transform infrared (FTIR) measurements (4000–400 cm^{-1}) were performed on a Spectrum Two FT-IR Spectrometer (PerkinElmer, Switzerland) with a universal ATR (UATR Two) cell equipped with a ZnSe single crystal. The acquisition performed using 60 scans and the resolution was set to 4 cm^{-1}. Zetasizer Nano ZEN5600 (Malvern, UK) was used to measure the zeta potential of the synthesized material. SU8030 (Hitachi, Japan) SEM-type microscope operated at an acceleration voltage of 10 kV and a probe current of 15 pA was used to examine the morphology of the material with scanning electron microscopy (SEM).

2.4. RSM with Box–Behnken Experimental Design

The influence of three independent operating parameters, i.e., catalyst concentration (A), phenol initial concentration (B), and pH of the solution (C), was considered in RSM. The remaining reaction conditions, namely, the airflow rate (50 mL/min) and reaction time (3 h), was kept constant in the experiment based on previous study [18]. The degradation efficiency of phenol (Equation (1)) was set as a response variable. Note that a previous study [18] was conducted to obtain the upper and lower limits of the parameters. Table 1 shows the ranges and levels of independent parameters A, B, and C. BBD was used to examine the combined effect of these three variables. Section 3.3 lists the set of experiments in table; it includes a replication of experiments at the central point. Regression analysis was the performed using OriginPro 2021 9.8.0.200 (OriginLab Corporation, Northampton, MA, USA) software. The suggested model's data were analyzed for significance and suitability using analysis for variance (ANOVA).

Table 1. Independent parameters and their ranges and levels.

Independent Parameters	Symbol	Range and Level		
		Low (−1)	Middle (0)	High (+1)
Catalyst concentration (g/L)	A	0.25	0.5	0.75
Phenol initial concentration (ppm)	B	20	60	100
pH	C	3	6.5	10

2.5. Photocatalytic Experiments

Figure 1 shows the photocatalytic experiments that were conducted in a jacketed glass reactor (working volume 225 mL) (Peschl Ultraviolet GmbH, Mainz, Germany) with a safety cabinet. The reactor was irradiated from inside using a custom-made LED immersion lamp; the LED has maximum emission at 430 nm. Glass reactor was then sonicated with a reaction mixture for uniform dispersion, followed by stirring with continuous airflow to maintain adsorption–desorption equilibrium for 30 min. Subsequently, lights were turned on, which is considered as zero time (t_o). Nine to ten samples (1 mL) were periodically collected from the reaction mixture. After centrifugation and filtration, the samples were analyzed using HPLC. For acidic and basic reaction conditions, the pH of the mixture was adjusted using 0.1 M HCl and NaOH. The phenol degradation efficiency was determined using the following Equation:

$$Degradation\ efficiency\ (\%) = \frac{C_o - C}{C_o} \times 100 \qquad (1)$$

where C_o is the initial phenol concentration and C is the residual phenol concentration in the solution at an irradiation time t.

Figure 1. Photocatalytic reactor setup.

The reduction of the reaction mixture volume due to the sampling was less than 5% at the end of the experiments and was therefore not considered in the calculation of the phenol degradation efficiency.

2.6. Analytical Techniques

A prominence HPLC system from Shimadzu (Kyoto, Japan) was used for analyzing the samples obtained from the reactor. The system is equipped with a binary pump (Model LC-20AB), an autosampler (Model SIL-20A), a degasser (Model DGU-20A3,) and a diode-array detector (Model SPD-M20A). Phenomenex (C18, 150 × 4.6 mm, 3 µm) column was used with a fixed flow rate of 0.8 mL/min, with the mobile phase gradient of water (A) and acetonitrile (B): starts with 15% B, followed by 60% B in 7 min and back to 15% B in 8 min;

injection of 5 µL; UV light of 254 nm. Phenol was analyzed at a maximum absorption wavelength (λmax) of 270 nm.

3. Results and Discussion
3.1. Photocatalyst Characterization

The metal-free g-C_3N_4 used in this study was synthesized and characterized in our previous study [18] using transmission electron microscopy (TEM), Brunauer–Emmett–Teller isotherms (BET), X-ray diffraction (XRD), X-ray photoelectron spectroscopy (XPS), photoluminescence (PL), and UV-Vis spectroscopy. In this study, scanning electron microscopy (SEM), Fourier transform infrared spectroscopy (FTIR), and zeta potential analyses were performed. Table 2 lists the physical properties of metal-free g-C_3N_4 before and after its exfoliation.

Table 2. Summary of characterization results [18].

Characterization	Bulk g-C_3N_4		Exfoliated g-C_3N_4	
BET	Surface area 11 m²/g	Pore size 1.91 Å	Surface area 170 m²/g	Pore size 1.96 Å
XRD	Weak peaks (2θ) 13.0°	Strong peaks (2θ) 27.2°	Weak peaks (2θ) 13.1°	Strong peaks (2θ) 27.4°
PL/UV-Vis	Max. absorption 458 nm	Bandgap 2.58 eV	Max. absorption 436 nm	Bandgap 2.68 eV
XPS	C1s peaks 288.2, 284.6, 286.2 and 292.9 eV	N1s peaks 398.5, 399.8, 400.8, 404.1 eV	C1s peaks 287.8, 284.7, 286.2 and 293.5 eV	N1s peaks 397.8, 399.1, 400.1, 403.5 eV

The exfoliated material has a significantly higher surface area than the bulk material, while the average pore size of both materials is almost the same (Table 2 and Figure S1). Using XRD, the material shows two characteristic peaks of g-C_3N_4 (Figure S4) [34,35]. The strong and weak peaks of N1s and C1s observed in XPS confirm the chemical state of g-C_3N_4 (Figure S3) [17,36–41]. Table 2 lists the maximum absorption wavelength and bandgap of the material, which are presented in Figure S2 [42,43].

In Figure 2, the selected SEM images of bulk and exfoliated g-C_3N_4 are presented. The thermal exfoliation transformed the stacked and aggregated structure of bulk g-C_3N_4 in a porous nanosheet structure. The reduction in layer thickness (Figure 2b) leads to an increase in the specific surface area of g-C_3N_4 [17,44–46].

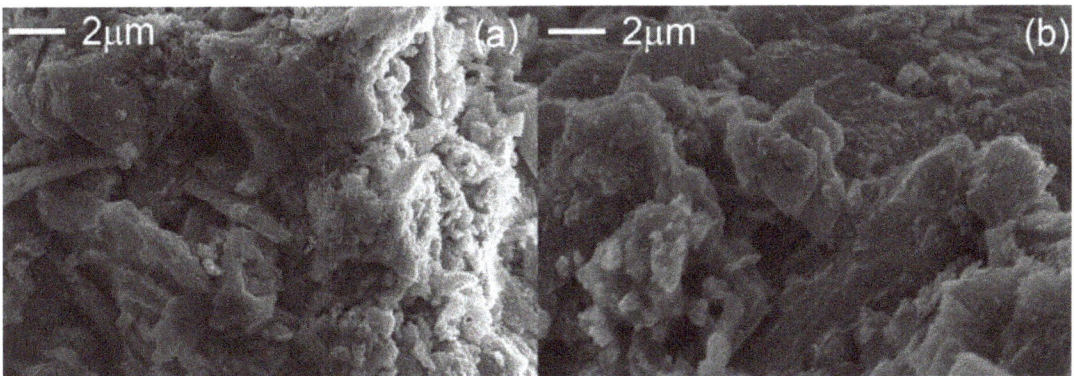

Figure 2. SEM images of the bulk (**a**) and exfoliated (**b**) g-C_3N_4.

Figure 3 shows the catalysts' FTIR spectra. A broad peak is observed between 3200 and 3000 cm^{-1}, which can be attributed to the stretching vibrations of N–H bonds from residual amino groups and adsorbed H_2O. The sharp peak that appears at 806 cm^{-1} can be attributed to the breathing mode of triazine units [47,48], whereas the strong bands between 1636 and 1242 cm^{-1} belong to the C=N and C–N bonds of heterocyclic rings. Because the spectra of both materials show the same absorption bands, the chemical structure remained unaltered after treatment.

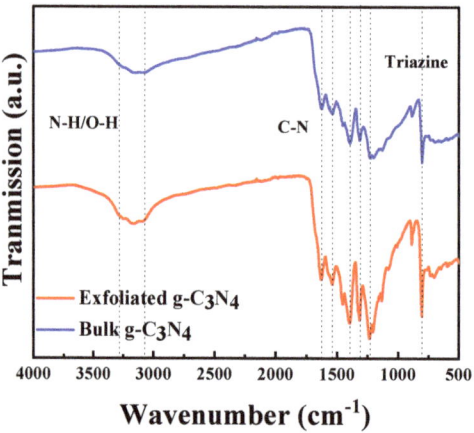

Figure 3. Fourier transform infrared spectra of bulk and exfoliated g-C_3N_4.

Figure 4 shows the effect of pH on the zeta potential of the exfoliated g-C_3N_4. The catalyst surface is positively charged at acidic pH (3) and negatively charged at natural (6) and basic pH (10).

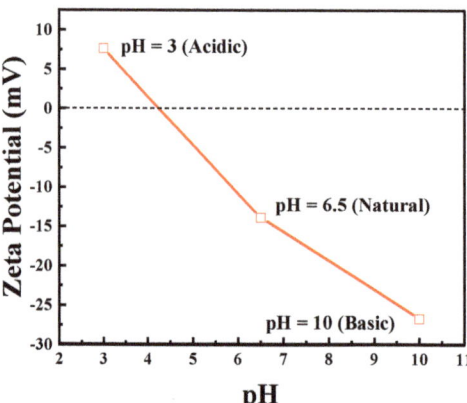

Figure 4. Zeta potential at different pH of the synthesized exfoliated g-C_3N_4. Reproduced with permission from [18].

The optical properties (PL/UV-Vis) and surface area (BET) of the material have changed with exfoliation; however, the chemical state (XPS), phase (XRD), and the chemical structure (FTIR) remained the same after exfoliation.

3.2. Photodegradation Studies

The photodegradation efficiency of exfoliated g-C_3N_4 photocatalyst was evaluated under visible light irradiation using 430 nm wavelength LEDs. The influence of individual operation parameters, catalyst concentration, phenol concentration, and pH of the solution, in their preselected ranges (Table 1), was examined. For all experiments, an adsorption time of 30 min was used before the light irradiation was started. Moreover, the photolysis experiment was performed to verify the removal of phenol in the absence of the catalyst. Phenol removal with adsorption in the dark and photolysis is insignificant compared to the removal of phenol obtained in the presence of light (Figure 5a). Figure 5a shows the effect of g-C_3N_4 photocatalyst concentration in the range of 0.1–0.75 g/L on phenol degradation, which increased with the increase in catalyst concentration up to 0.75 g/L because of an increased number of active sites available for the reaction to occur. However, there is no significant increase at >0.5 g/L because an additional increase of the catalyst concentration might cause light scattering and hindrance in light absorption. The effect of phenol concentration on the performance of the catalyst on phenol degradation was examined for three concentrations between 20 and 100 ppm and is shown in Figure 5b. The phenol degradation efficiency decreased as the concentration increased because of the higher number of molecules for adsorption on the available active sites, which hinders the absorption of light. Figure 5c shows the effect of different pH on phenol degradation. Increasing the pH decreases the degradation efficiency of exfoliated g-C_3N_4. Note that acidic pH is most favorable for phenol degradation because as per the zeta potential (Figure 3) and the surface charge of the catalyst is positive at an acidic pH, which helps attract OH^- ions produced in the solution due to dissociation of H_2O_2 to the surface and improves the degradation efficiency.

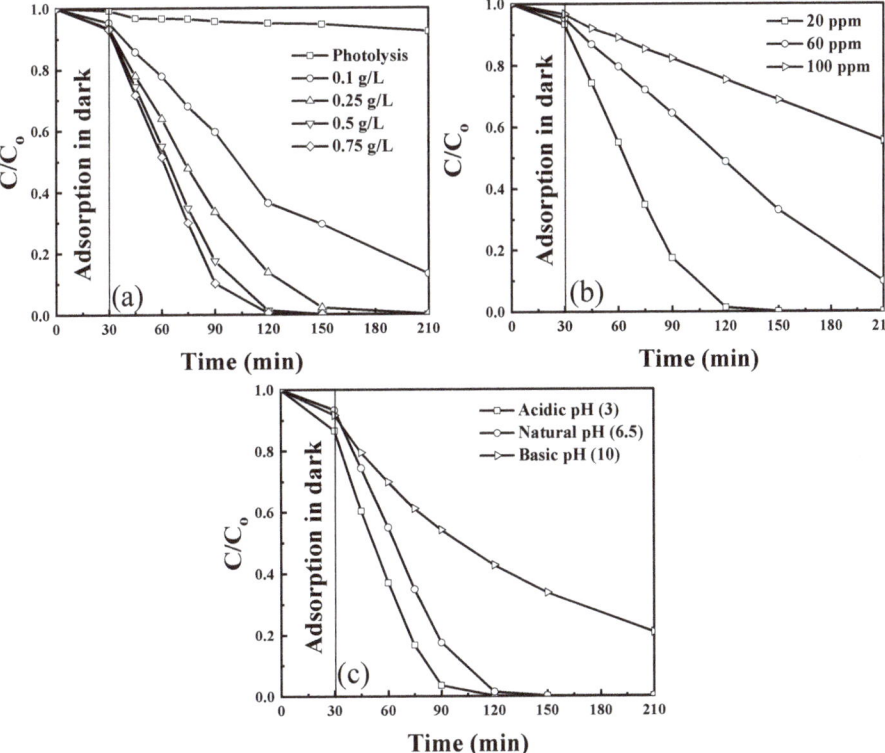

Figure 5. Phenol degradation at preselected (**a**) catalyst concentration (at 20 ppm and natural pH) (**b**) pollutant concentration (at 0.5 g/L and natural pH), and (**c**) pH of the solution (at 0.5 g/L and 20 ppm); airflow = 50 mL/min. Reproduced with permission from [18].

3.3. Response Surface Methodology
3.3.1. Model Equation

To analyze the combined effect of three variables: catalyst concentration (A), phenol concentration (B), and pH of the solution (C) on the degradation efficiency of phenol (Equation (1)), a three-variable BBD was used in the experimental design for RSM. Table 3 lists the set of performed experiments and the obtained phenol degradation (in 3 h and under an airflow of 50 mL/min).

Table 3. Box–Behnken design with experimental and predicted phenol degradation efficiency values with Equation (2).

Run	Experimental Conditions			Phenol Degradation Efficiency (%)	
	Catalyst Concentration (g/L)	Phenol Initial Concentration (ppm)	pH	Experimental	Predicted
1	0.25	100	6.5	43.49	44.23
2	0.50	60	6.5	82.25	85.72
3	0.25	20	6.5	100	93.95
4	0.75	60	3.0	94.09	86.18
5	0.75	20	6.5	100.00	100.00
6	0.50	20	10.0	79.18	74.07
7	0.50	20	3.0	100.00	100.00
8	0.50	60	6.5	84.93	85.72
9	0.25	60	10.0	40.77	43.02
10	0.50	60	6.5	85.94	85.72
11	0.50	100	10.0	24.09	24.35
12	0.50	100	3.0	54.43	54.79
13	0.75	60	10.0	53.15	55.74
14	0.25	60	3.0	70.39	73.46
15	0.50	60	6.5	88.37	85.72
16	0.75	100	6.5	58.31	56.95
17	0.50	60	6.5	87.12	85.72

Experimental data were fitted with four different models: two-factor interaction (2FI), linear, quadratic, and cubic model to obtain regression equations. Three different tests, namely, the sequential model sum of squares, lack of fit, and model summary statistics, were conducted to determine the adequacy of various models; the results are presented in Table 4. The response surface model is then used to select the best model based on the following criterion: the highest-order polynomial with additional significant terms and the model is not aliased (Table 4). The cubic model has the highest polynomial model because there are no sufficient unique design points to independently estimate all terms for that model. The aliased model results in unstable and inaccurate coefficients and graphs. Thus, the aliased model cannot be selected [49,50]. The criteria used in the lack of fit test is the non-significant lack of fit (p-value > 0.05) based on which a quadratic model is selected. Moreover, multiple summary statistics are calculated to compare models or to confirm the adequacy of the model. These statistics include adjusted R^2, predicted R^2, and prediction error sum of squares (PRESS). A good model will have a largely predicted r^2, and a low PRESS. According to the aforementioned criteria, adjusted R^2 (0.967) and predicted R^2 (0.805) are in reasonable agreement with each other and have a low PRESS. Thus, the quadratic model is finally selected to build the response surface.

Table 4. Adequacy of the models tested.

Source	Sum of Squares	Degree of Freedom	Mean Square	F Value	p-Value	Remark
		Sequential model sum of squares				
Linear	7118.98	3	2372.99	18.77	<0.0001	-
2FI	109.60	3	36.53	0.238	0.8678	-
Quadratic	1407.83	3	469.27	26.07	0.0004	Suggested
Cubic	104.30	3	34.76	6.41	0.0523	Aliased
		Lack of fit tests				
Linear	1621.73	9	180.19	33.22	<0.0021	-
2FI	1512.13	6	252.02	46.46	<0.0012	-
Quadratic	104.29	3	34.76	6.41	0.0523	Suggested
Cubic	0	0	-	-	-	Aliased

Source	Standard deviation	R^2	Adjusted R^2	Predicted R^2	PRESS	
		Model summary statistic				
Linear	11.24	0.8124	0.769	0.694	2678.06	-
2FI	12.38	0.8250	0.720	0.462	4712.53	-
Quadratic	4.24	0.9856	0.967	0.805	1702.70	Suggested
Cubic	2.33	0.9975	0.999	-	-	Aliased

Based on regression coefficients from Table 5, the following empirical second-order polynomial equation was obtained:

$$\text{Degradation Efficiency (\%)} = 85.72 + 6.36\,A - 24.86\,B - 15.22\,C + 3.71\,AB - 2.83\,AC - 2.38\,BC - 5.05\,A^2 - 5.22\,B^2 - 16.07\,C^2 \quad (2)$$

where, A, B, and C are the catalyst concentration, phenol concentration, and pH of the solution, respectively.

Table 5. Coefficients of the second-order polynomial (quadratic) equation.

Factor	Coefficient Estimate	Degree of Freedom	Standard Error	95% Confidence Interval Low	95% Confidence Interval Low	F Value	p-Value
Intercept	85.72	1	1.90	81.24	90.21	-	-
A	6.36	1	1.50	2.82	9.91	17.99	0.0038
B	−24.86	1	1.50	−28.40	−21.31	274.63	<0.0001
C	−15.22	1	1.50	−18.76	−11.67	102.89	<0.0001
AB	3.71	1	2.12	−1.31	8.72	3.05	0.1242
AC	−2.83	1	2.12	−7.85	2.19	1.78	0.2239
BC	−2.38	1	2.12	−7.40	2.64	1.26	0.2989
A^2	−5.05	1	2.07	−9.94	−0.16	5.96	0.0446
B^2	−5.22	1	2.07	−10.11	−0.33	6.38	0.0394
C^2	−16.07	1	2.07	−20.96	−11.18	60.44	0.0001

The influence of model terms on the degradation of phenol as per p-values (Table 5) is in the following order $B < C < C^2 < A < B^2 < A^2 < AB < AC < BC$. The mixed interaction terms AB, AC, and BC are not significant because their p–value is > 0.05 and may be removed from Equation (2).

An ANOVA of the second-order polynomial (Equation (2)) for phenol degradation was conducted; the results are shown in Table 6. In statistics, the significance of the model can be confirmed by a large F-value (53.31) and a small p-value (<0.0001). Furthermore, the significance of the model can be confirmed by the lack of fit test. In this study, the lack of fit is not significant because its p-value is >0.05. The accuracy of the model is confirmed by the low coefficient of variation (CV) value of 5.79%. The results showed that the signal-to-noise ratio of 24.89 is adequate.

Table 6. Analysis of variance ANOVA of the second-order polynomial (Equation (2)).

Source	Sum of Squares	Degree of Freedom	Mean Square	F Value	p-Value	Remark
Model	8636.42	9	959.60	53.31	<0.0001	Significant
Residual	126.00	7	18.00	-	-	-
Lack of fit	104.30	3	34.77	6.41	0.0523	Not Significant
Pure error	21.70	4	5.42	-	-	-
-	Adjusted R^2 = 0.967	Predicted R^2 = 0.810	Model precision = 24.89	-	-	-
-	Std. dev. = 4.24	Mean = 73.32	C.V. % = 5.79	-	-	-

Furthermore, the coefficient of determination R^2 confirmed the fit of the model. For the used model, the value of the predicted R^2 = 0.810 (Table 6) is in agreement with adjusted R^2 = 0.967, which indicates that the obtained model is significant.

Equation (2) provides a suitable relationship (R^2 = 0.810) between the response (degradation efficiency) and the parameters, which can be seen in Figure 6. In this figure, the experimental values of phenol degradation are plotted against the predicted values obtained from the RSM model; these values of the percentage phenol degradation fit well.

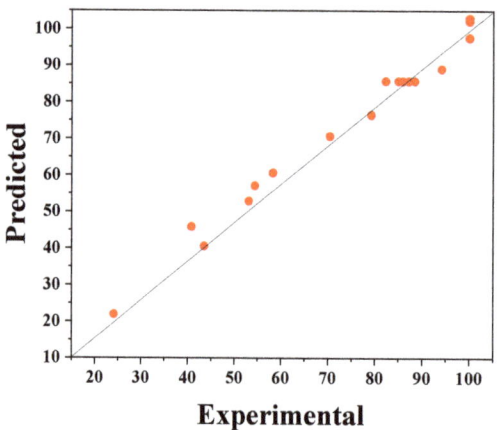

Figure 6. The experimental phenol degradation efficiency (%) plotted against the predicted values from the RSM model.

3.3.2. Interaction Effects of Independent Operating Parameters

Three dimensional (3D) response surface and contour plots were generated using the regression model (Equation (2)) to visualize the influence of the independent operating parameters on phenol degradation; they are presented in Figures 7–9. In surface and contour plots, one parameter is maintained constant at its zero levels, whereas the other two are varied in the studied range reported in Table 1.

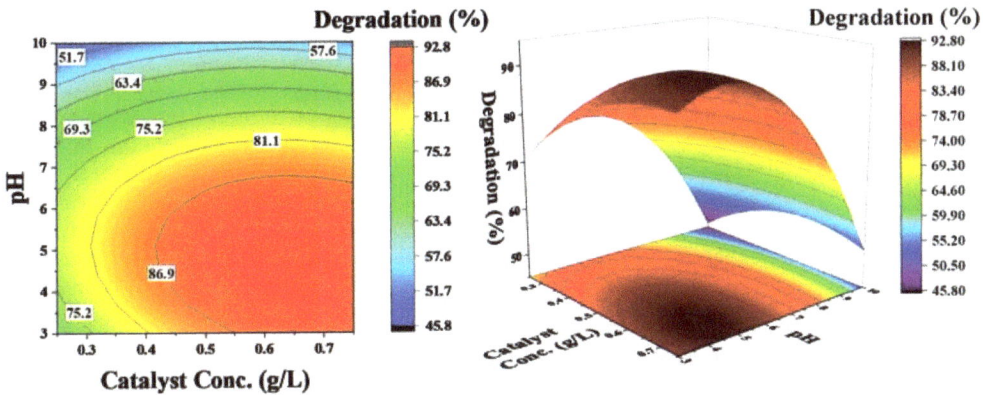

Figure 7. Effect of catalyst concentration and pH on the degradation of phenol: pollutant concentration was kept constant at 60 ppm.

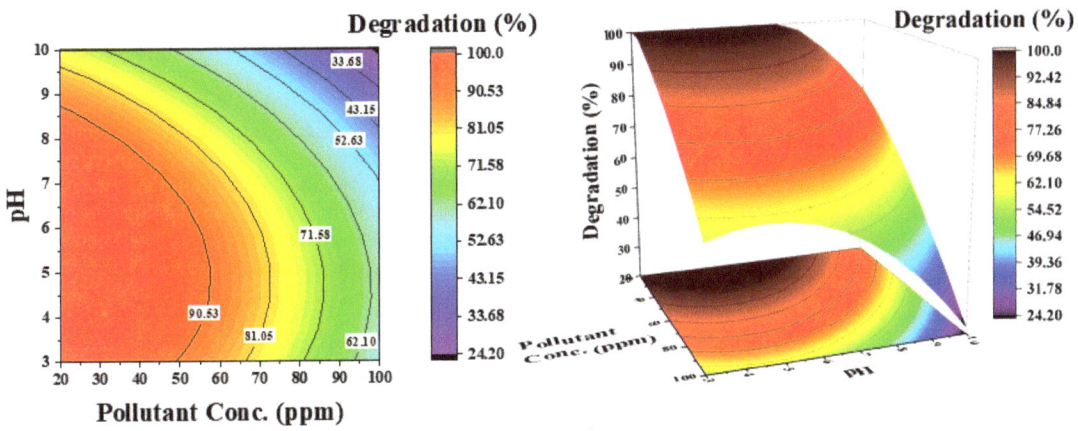

Figure 8. Effect of pollutant concentration and pH on the degradation of phenol: catalyst concentration was kept constant at 0.5 g/L.

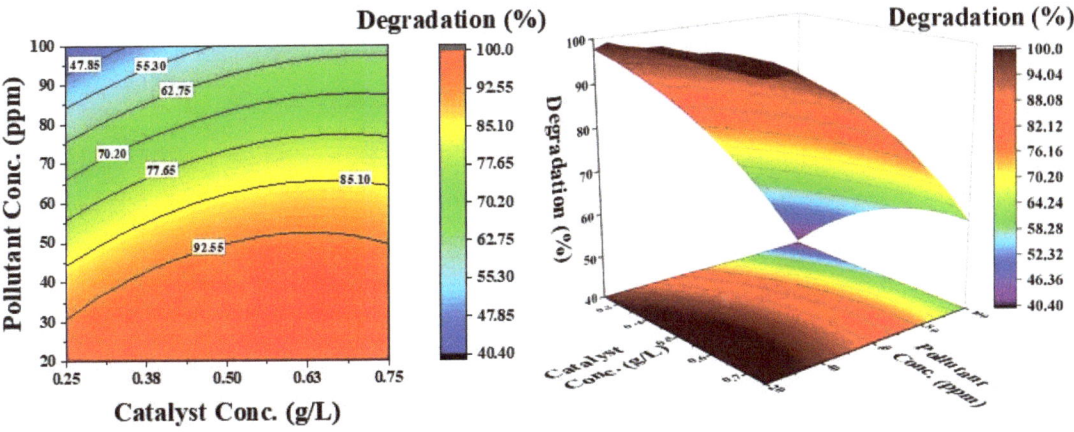

Figure 9. Effect of catalyst concentration and pollutant concentration on the degradation of phenol: pH was kept constant at 6.5.

Figure 7 shows the influence of pH and catalyst concentration on the degradation efficiency of phenol at a constant phenol concentration of 60 ppm. The contour lines show a decrease in the degradation efficiency with an increase in pH; there is no considerable increase in efficiency, even at higher catalyst concentrations. However, an increase in degradation efficiency with a decrease in pH is observed. These results demonstrate that pH has a significant effect on phenol degradation and a low pH favors the degradation process. This phenomenon is linked with the zeta potential of the catalyst surface [18]. There is a positive charge at the surface of the catalyst at an acidic pH (Figure 2), which attracts the OH^- ions produced in the solution due to dissociation of H_2O_2 and significantly increases the degradation process. However, at a basic pH, the surface charge is negative and there could be electrostatic repulsion that reduces the efficiency of the degradation process.

Figure 8 shows the influence of pH and pollutant concentration on phenol degradation at a constant catalyst concentration of 0.5 g/L. For selecting the catalyst concentration, the effect of initial pollutant concentration is important. The contour lines demonstrate that simultaneously increasing both parameters (pH and phenol concentration) considerably decreases the degradation efficiency of phenol (33%), which is 62% at a low pH. As shown in Figure 5b, at low pH and low pollutant concentration, 100% degradation is achieved in a considered reaction time of 3 h. An increase in degradation efficiency from high to low pH can then be associated with catalyst surface charge. However, a decrease in efficiency at low pH from low to high phenol concentration is attributed to the increased number of pollutant molecules compared with the available active sites.

Figure 9 shows the effect of catalyst concentration and pollutant concentration at a constant pH of 6.5. The contour lines demonstrate that both parameters independently affect the degradation efficiency. By increasing the catalyst concentration at a lower pollutant concentration, phenol degradation increases; however, at a higher pollutant concentration, the degradation efficiency decreases. This can be attributed to the availability of active sites on the catalyst surface for OH^- radicals, as well as phenol molecules. The electron–hole pair generated from the catalyst surface improves the degradation rate.

3.3.3. Experimental Validation of RSM Model

To demonstrate the applicability of the model, a hypothetical case study for water with a phenol concentration of 50 ppm was considered. The model equation was used to identify the optimum catalyst concentration and pH, leading to maximal phenol degradation in 3 h under an airflow rate of 50 mL/min. According to the model prediction, maximal phenol degradation of 88.62% is achievable using 0.4 g/L of catalyst concentration and operating at a pH of 6.5. To examine the accuracy of the model prediction, an experiment was conducted under these conditions. The experimentally obtained phenol degradation was 83.75%, which is less than a 5% deviation from the predicted value. Thus, the optimum operating point obtained by RSM was successfully confirmed; this suggests that RSM can be a useful tool for optimizing photocatalytic processes. Similarly, the model developed can be used for minimizing the catalyst amount or for maximizing the degradation efficiency of phenols for any set of parameters in range.

4. Conclusions

Metal-free $g-C_3N_4$ was used for the photocatalytic degradation of phenol from an aqueous solution. The morphology of the catalyst was confirmed by SEM, and the surface charge was confirmed using zeta potential. Based on zeta potential, the catalyst surface was confirmed to have a positive surface charge under acidic conditions and a negative surface charge under basic conditions; therefore, acidic pH favors the degradation process. A RSM based on the BBD was used to analyze the degradation efficiency of phenol. The influence of experimental parameters, namely, catalyst concentration, pollutant concentration, and pH of the solution, and their interaction at a different level was examined for phenol degradation. An empirical regression quadratic model was developed for the response variable. Analysis of variance (ANOVA) demonstrated that the model is significant with an insignificant lack of fit and a high coefficient of determination (R^2) of 0.96, which can be helpful to navigate the design space. Furthermore, an optimized degradation efficiency of 83.75% was achieved for phenol concentration of 50 ppm, catalyst concentration of 0.4 g/L, and a solution pH of 6.5 pH (in 3 h and under an airflow of 50 mL/min). Thus, the results suggest that the RSM can be used for the optimization of parameters for maximizing the photocatalytic degradation of phenol using g-C_3N_4 and LEDs.

Supplementary Materials: The following are available online at https://www.mdpi.com/article/10.3390/catal11080898/s1, Figure S1 N_2 adsorption-desorption isotherms of bulk and exfoliated g-C_3N_4. The inset shows the corresponding BJH pore size distribution curves of the sample, Figure S2 (**a**) UV-Vis absorption spectra and (**b**) PL spectra of bulk and exfoliated g-C_3N_4; insets of (**a**) showing the Tauc plots, Figure S3 XPS spectra of bulk and exfoliated g-C_3N_4 C1s, N1s, Figure S4 X-ray diffraction patterns of bulk and exfoliated g-C_3N_4.

Author Contributions: Conceptualization, A.G.R.; Formal analysis, A.G.R.; Investigation, A.G.R.; Methodology, A.G.R.; Resources, M.M.; Supervision, M.M.; Writing—original draft, A.G.R.; Writing—review and editing, M.M. All authors have read and agreed to the published version of the manuscript.

Funding: This research received no external funding.

Acknowledgments: A.G.R. acknowledges the financial support from the Higher Education Commission, Pakistan, and Deutscher Akademischer Austauschdienst (DAAD), Germany.

Conflicts of Interest: The authors declare no conflict of interest.

References

1. Zulfiqar, M.; Samsudin, M.F.R.; Sufian, S. Modelling and optimization of photocatalytic degradation of phenol via TiO_2 nanoparticles: An insight into response surface methodology and artificial neural network. *J. Photochem. Photobiol. A Chem.* **2019**, *384*, 112039. [CrossRef]
2. Jourshabani, M.; Shariatinia, Z.; Badiei, A. Facile one-pot synthesis of cerium oxide/sulfur-doped graphitic carbon nitride (g-C_3N_4) as efficient nanophotocatalysts under visible light irradiation. *J. Colloid Interface Sci.* **2017**, *507*, 59–73. [CrossRef] [PubMed]
3. Hassani, A.; Eghbali, P.; Metin, O. Sonocatalytic removal of methylene blue from water solution by cobalt ferrite/mesoporous graphitic carbon nitride ($CoFe_2O_4$/mpg-C_3N_4) nanocomposites: Response surface methodology approach. *Environ. Sci. Pollut. Res. Int.* **2018**, *25*, 32140–32155. [CrossRef] [PubMed]
4. Mirzaei, A.; Yerushalmi, L.; Chen, Z.; Haghighat, F. Photocatalytic degradation of sulfamethoxazole by hierarchical magnetic ZnO@g-C_3N_4: RSM optimization, kinetic study, reaction pathway and toxicity evaluation. *J. Hazard. Mater.* **2018**, *359*, 516–526. [CrossRef]
5. Choquette-Labbé, M.; Shewa, W.; Lalman, J.; Shanmugam, S. Photocatalytic Degradation of Phenol and Phenol Derivatives Using a Nano-TiO_2 Catalyst: Integrating Quantitative and Qualitative Factors Using Response Surface Methodology. *Water* **2014**, *6*, 1785–1806. [CrossRef]
6. Yasar Arafath, K.A.; Baskaralingam, P.; Gopinath, S.; Nilavunesan, D.; Sivanesan, S. Degradation of phenol from retting-pond wastewater using anaerobic sludge reactor integrated with photo catalytic treatment. *Chem. Phys. Lett.* **2019**, *734*, 136727. [CrossRef]
7. Hararah, M.A.; Ibrahim, K.A.; Al-Muhtaseb, A.a.H.; Yousef, R.I.; Abu-Surrah, A.; Qatatsheh, A.a. Removal of phenol from aqueous solutions by adsorption onto polymeric adsorbents. *J. Appl. Polym. Sci.* **2010**, *117*, 1908–1913. [CrossRef]
8. Aslam, Z.; Qaiser, M.; Ali, R.; Abbas, A.; Ihsanullah; Zarin, S. Al_2O_3/MnO_2/CNTs nanocomposite: Synthesis, characterization and phenol adsorption. *Fuller. Nanotub. Carbon Nanostruct.* **2019**, *27*, 591–600. [CrossRef]
9. Huang, C.H.; Liou, R.M.; Chen, S.H.; Hung, M.Y.; Lai, C.L.; Lai, J.Y. Microbial degradation of phenol in a modified three-stage airlift packing-bed reactor. *Water Environ. Res.* **2010**, *82*, 249–258. [CrossRef]
10. Yavuz, Y.; Savas Koparal, A.; Bakir Öğütveren, Ü. Phenol Removal through Chemical Oxidation using Fenton Reagent. *Chem. Eng. Technol.* **2007**, *30*, 583–586. [CrossRef]
11. Asanjarani, N.; Bagtash, M.; Zolgharnein, J. A comparison between Box–Behnken design and artificial neural network: Modeling of removal of Phenol Red from water solutions by nanocobalt hydroxide. *J. Chemom.* **2020**, *34*, e3283. [CrossRef]
12. Peng, H.; Zou, C.; Wang, C.; Tang, W.; Zhou, J. The effective removal of phenol from aqueous solution via adsorption on CS/beta-CD/CTA multicomponent adsorbent and its application for COD degradation of drilling wastewater. *Environ. Sci. Pollut. Res. Int.* **2020**, *27*, 33668–33680. [CrossRef] [PubMed]
13. Chowdhury, P.; Nag, S.; Ray, A.K. Degradation of Phenolic Compounds through UV and Visible-Light-Driven Photocatalysis: Technical and Economic Aspects. In *Phenolic Compounds—Natural Sources, Importance and Applications*; IntechOpen: London, UK, 2017. [CrossRef]

14. Tetteh, E.K.; Rathilal, S.; Naidoo, D.B. Photocatalytic degradation of oily waste and phenol from a local South Africa oil refinery wastewater using response methodology. *Sci. Rep.* **2020**, *10*, 8850. [CrossRef] [PubMed]
15. Md Rosli, N.I.; Lam, S.-M.; Sin, J.-C.; Satoshi, I.; Mohamed, A.R. Photocatalytic Performance of ZnO/g-C_3N_4 for Removal of Phenol under Simulated Sunlight Irradiation. *J. Environ. Eng.* **2018**, *144*, 04017091. [CrossRef]
16. Rana, A.G.; Ahmad, W.; Al-Matar, A.; Shawabkeh, R.; Aslam, Z. Synthesis and characterization of Cu-Zn/TiO_2 for the photocatalytic conversion of CO_2 to methane. *Environ. Technol.* **2017**, *38*, 1085–1092. [CrossRef]
17. Lima, M.J.; Silva, A.M.T.; Silva, C.G.; Faria, J.L. Graphitic carbon nitride modified by thermal, chemical and mechanical processes as metal-free photocatalyst for the selective synthesis of benzaldehyde from benzyl alcohol. *J. Catal.* **2017**, *353*, 44–53. [CrossRef]
18. Rana, A.G.; Tasbihi, M.; Schwarze, M.; Minceva, M. Efficient Advanced Oxidation Process (AOP) for Photocatalytic Contaminant Degradation Using Exfoliated Metal-Free Graphitic Carbon Nitride and Visible Light-Emitting Diodes. *Catalysts* **2021**, *11*, 662. [CrossRef]
19. Al-Kandari, H.; Abdullah, A.M.; Ahmad, Y.H.; Al-Kandari, S.; AlQaradawi, S.Y.; Mohamed, A.M. An efficient eco advanced oxidation process for phenol mineralization using a 2D/3D nanocomposite photocatalyst and visible light irradiations. *Sci. Rep.* **2017**, *7*, 9898. [CrossRef] [PubMed]
20. Moradi, V.; Ahmed, F.; Jun, M.B.G.; Blackburn, A.; Herring, R.A. Acid-treated Fe-doped TiO_2 as a high performance photocatalyst used for degradation of phenol under visible light irradiation. *J. Environ. Sci.* **2019**, *83*, 183–194. [CrossRef] [PubMed]
21. Nobijari, L.A.; Schwarze, M.; Tasbihi, M. Photocatalytic Degradation of Phenol Using Photodeposited Pt Nanoparticles on Titania. *J. Nanosci. Nanotechnol.* **2020**, *20*, 1056–1065. [CrossRef] [PubMed]
22. Jourshabani, M.; Shariatinia, Z.; Badiei, A. In situ fabrication of SnO_2/S-doped g-C_3N_4 nanocomposites and improved visible light driven photodegradation of methylene blue. *J. Mol. Liq.* **2017**, *248*, 688–702. [CrossRef]
23. Jourshabani, M.; Shariatinia, Z.; Badiei, A. Sulfur-Doped Mesoporous Carbon Nitride Decorated with Cu Particles for Efficient Photocatalytic Degradation under Visible-Light Irradiation. *J. Phys. Chem. C* **2017**, *121*, 19239–19253. [CrossRef]
24. Jourshabani, M.; Shariatinia, Z.; Badiei, A. Controllable Synthesis of Mesoporous Sulfur-Doped Carbon Nitride Materials for Enhanced Visible Light Photocatalytic Degradation. *Langmuir* **2017**, *33*, 7062–7078. [CrossRef]
25. Lee, S.C.; Lintang, H.O.; Yuliati, L. A urea precursor to synthesize carbon nitride with mesoporosity for enhanced activity in the photocatalytic removal of phenol. *Chem. Asian J.* **2012**, *7*, 2139–2144. [CrossRef] [PubMed]
26. Ren, H.-T.; Jia, S.-Y.; Wu, Y.; Wu, S.-H.; Zhang, T.-H.; Han, X. Improved Photochemical Reactivities of Ag_2O/g-C_3N_4 in Phenol Degradation under UV and Visible Light. *Ind. Eng. Chem. Res.* **2014**, *53*, 17645–17653. [CrossRef]
27. Deng, P.; Gan, M.; Zhang, X.; Li, Z.; Hou, Y. Non-noble-metal Ni nanoparticles modified N-doped g-C_3N_4 for efficient photocatalytic hydrogen evolution. *Int. J. Hydrogen Energy* **2019**, *44*, 30084–30092. [CrossRef]
28. Hu, J.Y.; Tian, K.; Jiang, H. Improvement of phenol photodegradation efficiency by a combined g-C_3N_4/Fe(III)/persulfate system. *Chemosphere* **2016**, *148*, 34–40. [CrossRef] [PubMed]
29. Huang, Z.; Li, F.; Chen, B.; Lu, T.; Yuan, Y.; Yuan, G. Well-dispersed g-C_3N_4 nanophases in mesoporous silica channels and their catalytic activity for carbon dioxide activation and conversion. *Appl. Catal. B Environ.* **2013**, *136–137*, 269–277. [CrossRef]
30. Sharma, M.; Vaidya, S.; Ganguli, A.K. Enhanced photocatalytic activity of g-C_3N_4-TiO_2 nanocomposites for degradation of Rhodamine B dye. *J. Photochem. Photobiol. A Chem.* **2017**, *335*, 287–293. [CrossRef]
31. Hernández-Uresti, D.B.; Vázquez, A.; Sanchez-Martinez, D.; Obregón, S. Performance of the polymeric g-C_3N_4 photocatalyst through the degradation of pharmaceutical pollutants under UV–vis irradiation. *J. Photochem. Photobiol. A Chem.* **2016**, *324*, 47–52. [CrossRef]
32. Abdullah, A.H.; Moey, H.J.M.; Yusof, N.A. Response surface methodology analysis of the photocatalytic removal of Methylene Blue using bismuth vanadate prepared via polyol route. *J. Environ. Sci.* **2012**, *24*, 1694–1701. [CrossRef]
33. Song, C.; Li, X.; Wang, L.; Shi, W. Fabrication, Characterization and Response Surface Method (RSM) Optimization for Tetracycline Photodegration by $Bi_{3.84}W_{0.16}O_{6.24}$-graphene oxide (BWO-GO). *Sci. Rep.* **2016**, *6*, 37466. [CrossRef]
34. Wang, J.; Guo, P.; Dou, M.; Wang, J.; Cheng, Y.; Jönsson, P.G.; Zhao, Z. Visible light-driven g-C_3N_4/m-$Ag_2Mo_2O_7$ composite photocatalysts: Synthesis, enhanced activity and photocatalytic mechanism. *RSC Adv.* **2014**, *4*, 51008–51015. [CrossRef]
35. Wang, Y.; Yang, W.; Chen, X.; Wang, J.; Zhu, Y. Photocatalytic activity enhancement of core-shell structure g-C_3N_4@TiO_2 via controlled ultrathin g-C_3N_4 layer. *Appl. Catal. B Environ.* **2018**, *220*, 337–347. [CrossRef]
36. Yuan, Y.-J.; Shen, Z.; Wu, S.; Su, Y.; Pei, L.; Ji, Z.; Ding, M.; Bai, W.; Chen, Y.; Yu, Z.-T.; et al. Liquid exfoliation of g-C_3N_4 nanosheets to construct 2D-2D MoS_2/g-C_3N_4 photocatalyst for enhanced photocatalytic H_2 production activity. *Appl. Catal. B Environ.* **2019**, *246*, 120–128. [CrossRef]
37. Muñoz-Batista, M.J.; Rodríguez-Padrón, D.; Puente-Santiago, A.R.; Kubacka, A.; Luque, R.; Fernández-García, M. Sunlight-Driven Hydrogen Production Using an Annular Flow Photoreactor and g-C_3N_4-Based Catalysts. *ChemPhotoChem* **2018**, *2*, 870–877. [CrossRef]
38. Yuan, X.; Zhou, C.; Jin, Y.; Jing, Q.; Yang, Y.; Shen, X.; Tang, Q.; Mu, Y.; Du, A.K. Facile synthesis of 3D porous thermally exfoliated g-C_3N_4 nanosheet with enhanced photocatalytic degradation of organic dye. *J. Colloid Interface Sci.* **2016**, *468*, 211–219. [CrossRef]
39. Xu, J.; Zhang, L.; Shi, R.; Zhu, Y. Chemical exfoliation of graphitic carbon nitride for efficient heterogeneous photocatalysis. *J. Mater. Chem. A* **2013**, *1*, 14766–14772. [CrossRef]
40. Li, Y.; Wang, M.-Q.; Bao, S.-J.; Lu, S.; Xu, M.; Long, D.; Pu, S. Tuning and thermal exfoliation graphene-like carbon nitride nanosheets for superior photocatalytic activity. *Ceram. Int.* **2016**, *42*, 18521–18528. [CrossRef]

41. Yang, L.; Liu, X.; Liu, Z.; Wang, C.; Liu, G.; Li, Q.; Feng, X. Enhanced photocatalytic activity of g-C_3N_4 2D nanosheets through thermal exfoliation using dicyandiamide as precursor. *Ceram. Int.* **2018**, *44*, 20613–20619. [CrossRef]
42. Papailias, I.; Giannakopoulou, T.; Todorova, N.; Demotikali, D.; Vaimakis, T.; Trapalis, C. Effect of processing temperature on structure and photocatalytic properties of g-C_3N_4. *Appl. Surf. Sci.* **2015**, *358*, 278–286. [CrossRef]
43. Yu, B.; Meng, F.; Khan, M.W.; Qin, R.; Liu, X. Facile synthesis of AgNPs modified TiO_2@g-C_3N_4 heterojunction composites with enhanced photocatalytic activity under simulated sunlight. *Mater. Res. Bull.* **2020**, *121*, 110641. [CrossRef]
44. Yang, Y.; Lei, W.; Xu, Y.; Zhou, T.; Xia, M.; Hao, Q. Determination of trace uric acid in serum using porous graphitic carbon nitride (g-C3N4) as a fluorescent probe. *Microchim. Acta* **2017**, *185*, 39. [CrossRef] [PubMed]
45. Dong, F.; Wang, Z.; Sun, Y.; Ho, W.K.; Zhang, H. Engineering the nanoarchitecture and texture of polymeric carbon nitride semiconductor for enhanced visible light photocatalytic activity. *J. Colloid Interface Sci.* **2013**, *401*, 70–79. [CrossRef] [PubMed]
46. Sturini, M.; Speltini, A.; Maraschi, F.; Vinci, G.; Profumo, A.; Pretali, L.; Albini, A.; Malavasi, L. g-C3N4-promoted degradation of ofloxacin antibiotic in natural waters under simulated sunlight. *Environ. Sci. Pollut. Res. Int.* **2017**, *24*, 4153–4161. [CrossRef] [PubMed]
47. Zhu, B.; Xia, P.; Ho, W.; Yu, J. Isoelectric point and adsorption activity of porous g-C_3N_4. *Appl. Surf. Sci.* **2015**, *344*, 188–195. [CrossRef]
48. Praus, P.; Svoboda, L.; Dvorský, R.; Reli, M. Nanocomposites of SnO2 and g-C_3N_4: Preparation, characterization and photocatalysis under visible LED irradiation. *Ceram. Int.* **2018**, *44*, 3837–3846. [CrossRef]
49. Kumar, M.; Ponselvan, F.I.; Malviya, J.R.; Srivastava, V.C.; Mall, I.D. Treatment of bio-digester effluent by electrocoagulation using iron electrodes. *J. Hazard. Mater.* **2009**, *165*, 345–352. [CrossRef] [PubMed]
50. Liu, J.; Wang, J.; Leung, C.; Gao, F. A Multi-Parameter Optimization Model for the Evaluation of Shale Gas Recovery Enhancement. *Energies* **2018**, *11*, 654. [CrossRef]

Article

On the Role of the Cathode for the Electro-Oxidation of Perfluorooctanoic Acid

Alicia L. Garcia-Costa [1,2,*], Andre Savall [2], Juan A. Zazo [1], Jose A. Casas [1] and Karine Groenen Serrano [2,*]

1. Chemical Engineering Department, Universidad Autónoma de Madrid, 28049 Madrid, Spain; juan.zazo@uam.es (J.A.Z.); jose.casas@uam.es (J.A.C.)
2. Laboratoire de Génie Chimique, Université de Toulouse, CNRS, INPT, UPS, 31062 Toulouse CEDEX 9, France; savall@chimie.ups-tlse.fr
* Correspondence: alicial.garcia@uam.es (A.L.G.C.); serrano@chimie.ups-tlse.fr (K.G.S.)

Received: 22 July 2020; Accepted: 6 August 2020; Published: 8 August 2020

Abstract: Perfluorooctanoic acid (PFOA), $C_7F_{15}COOH$, has been widely employed over the past fifty years, causing an environmental problem because of its dispersion and low biodegradability. Furthermore, the high stability of this molecule, conferred by the high strength of the C-F bond makes it very difficult to remove. In this work, electrochemical techniques are applied for PFOA degradation in order to study the influence of the cathode on defluorination. For this purpose, boron-doped diamond (BDD), Pt, Zr, and stainless steel have been tested as cathodes working with BDD anode at low electrolyte concentration (3.5 mM) to degrade PFOA at 100 mg/L. Among these cathodic materials, Pt improves the defluorination reaction. The electro-degradation of a PFOA molecule starts by a direct exchange of one electron at the anode and then follows a complex mechanism involving reaction with hydroxyl radicals and adsorbed hydrogen on the cathode. It is assumed that Pt acts as an electrocatalyst, enhancing PFOA defluorination by the reduction reaction of perfluorinated carbonyl intermediates on the cathode. The defluorinated intermediates are then more easily oxidized by HO• radicals. Hence, high mineralization (x_{TOC}: 76.1%) and defluorination degrees (x_F^-: 58.6%) were reached with Pt working at current density $j = 7.9$ mA/cm^2. This BDD-Pt system reaches a higher efficiency in terms of defluorination for a given electrical charge than previous works reported in literature. Influence of the electrolyte composition and initial pH are also explored.

Keywords: perfluorooctanoic acid; emerging contaminant; defluorination; platinum; electro-oxidation

1. Introduction

Perfluoroalkyl substances (PFAS), such as perfluorooctanoic acid (PFOA, $C_7F_{15}COOH$) are widely used in the chemical industry because of their amphiphilicity, stability, and surfactant property. They are employed in the synthesis of fluoropolymers and fluoroelastomers, as surfactants in fire-fighting foams, and in textile and paper industries to produce water and oil repellent surfaces [1]. Nevertheless, despite their practical interest, these substances present a high toxicity due to their potential bioaccumulation, and common occurrence in water resources. PFOA has been recognized as an emerging environmental pollutant and has been included in the European Candidate List of Substances of Very High Concern ("SVHC") [2]. Hence, the current challenge is to develop highly efficient and cost-effective processes for the elimination of perfluoroalkyl substances at source.

The main issue in PFOA degradation is to break the C-F bond, one of the strongest bonds known ($\approx$460 kJ/mol) [3]. This confers a high stability and resistance to PFAS which cannot be degraded by direct hydrolysis, photolysis, or through conventional biological treatments [4]. As a result, PFAS have been detected in natural water streams [5], sediments [6], and even in tap and bottled water in concentrations up to 640 ng/L [7]. So far, adsorption onto carbonaceous materials [8], alumina [9], or other sorbents [10] have been successfully applied for PFAS removal. Nonetheless, this technology implies the transfer of the pollutant to another phase, the sorbent, which becomes a new residue after use. To overcome this drawback, advanced oxidation processes (AOP) are being explored for PFAS removal. AOP are based on the use of strong oxidizing radicals to degrade, most commonly, organic pollutants in aqueous phase [11]. The most extended AOP are those based on the use of hydroxyl radicals (HO•) to attack organic pollutants by hydrogen abstraction [12]. Consequently, the substitution of all organic hydrogen for fluorine in PFOA makes these compounds inert to this kind of AOP. The non-reactivity of PFOA to HO• attack has been confirmed by various studies [13–15]. As a matter of fact, Maruthamuthu et al. have shown that the reactivity of hydroxyl radicals on acetate decreases considerably with increasing halogen substitution [13]. Using the Fenton process, known to generate hydroxyl radicals by the action of Fe (II) on hydrogen peroxide, no degradation was observed when the Fenton reagent (0.2 mM, Fe^{2+}: H_2O_2, molar ratio = 1:1) was mixed with PFOA (0.02 mM) at room temperature [14]. Similar results were obtained by Santos et al. with only 10% PFOA removal and without any C-F bond cleavage [16].

More recently, photocatalytic treatments have been applied for PFOA degradation. This technology achieved high PFOA removal (x_{PFOA} > 90%) when using modified TiO_2 photocatalysts such as Cu-TiO_2 [17], Pb-TiO_2 [18], and rGO-TiO_2 [19].

Besides PFOA removal, defluorination (x_{F2212^-}) is a very important parameter to evaluate the process efficiency. x_F^- defined as the ratio of the fluoride concentration ($C_{F^-,\ measured}$) released by PFOA degradation with respect to the initial content of fluoride in the initial amount of PFOA molecule ($C_{F,PFOA\ 0}$) is expressed in percentage as shown in Equation (1).

$$x_{F^-} = \frac{C_{F^-,\ measured}}{C_{F,\ PFOA\ 0}} \cdot 100 \qquad (1)$$

In photochemical oxidation of PFOA, defluorination is usually low (x_F^- < 25%), with the average x_F^-/x_{PFOA} ratio around 0.26 [20].

Another technique for PFAS remediation is electrochemical degradation. PFOA electrooxidation has been successfully carried out in different systems using boron-doped diamond (BDD) as anode (Table 1). Under the studied conditions, PFOA removal ranged from 60% to 100%. It should be noted that defluorination values were very different, suggesting that either the operating conditions (electrolyte, pH, etc.,) or the cathode reduction reactions may play a key role in the PFOA degradation mechanism. The cleavage of the C-F bonds to form F^- ions is interesting because F^- ions readily combine with Ca^{2+} to form environmentally harmless CaF_2, as reported by Hori et al. [3].

x_F^- and x_{PFOA} ratios obtained by electrochemical treatment (up to 80–85%, Table 1) are higher than those reported in photo-oxidation (<25%). Nonetheless, all previous electrooxidation studies were conducted employing a high supporting electrolyte concentration, which makes difficult to dispose the treated wastewater after reaction. Therefore, this work aims to gain knowledge on the role of the cathode as electrocatalyst in PFOA electrooxidation working at low electrolyte concentration (3.5 mM). For this purpose, BDD was chosen as the anode and BDD, Pt, Zr, and stainless steel were tested as cathodes in the degradation of 100 mg/L PFOA.

Table 1. Perfluorooctanoic acid (PFOA) electrooxidation with boron-doped diamond (BDD) anode.

Electro-Oxidation System	Operating Conditions	Results	Remarks	Ref
Anode: BDD Cathode: W Area: 38 cm^2 Spacing: 4 mm	[PFOA]$_0$: 15 mg/L 8Electrolyte: 1500 mg/L Na$_2$SO$_4$ or 1500 mg/L Na$_2$SO$_4$ + 167 mg/L NaCl T: 20 °C, j: 3, 15, 50 mA/cm^2 V: 250 mL, t: 480 min	x_{PFOA}: 60–100% x_F^-: 30–80%	No apparent influence of Cl$^-$ in the process	[21]
Anode: BDD Cathode: W Area: 42 cm^2 Spacing: 8 mm	[PFOA]$_0$: 100 mg/L Electrolyte: 1.4–8.4 g/L NaClO$_4$, 5 g/L NaSO$_4$ T: 20 °C, j: 50, 100, 200 mA/cm^2 V: not specified, t: 360 min	x_{PFOA}: 93% x_{TOC}: 95% x_F^-: 38%	In the tested conditions, SO$_4^{2-}$ did not produce additional oxidants. higher j, higher degradation	[22]
Anode: BDD Cathode: BDD Area: 85 cm^2 Spacing: 30 mm	[PFOA]$_0$: 50 mg/L Electrolyte: 1.4 g/L NaClO$_4$ pH: 3, 9, 12, T: 32 °C, j: 23.24 mA/cm^2 V: 40 mL, t: 120 min	x_{PFOA}: 100% x_F^-: 58%	Slightly better results obtained at pH$_0$ 3 than pH$_0$ 9	[23]
Anode: BDD Cathode: Pt Area: 77.4 cm^2 Spacing: 10 mm	[PFOA]$_0$: 50 mg/L Electrolyte: 1.2 g/L NaClO$_4$ T: not specified, j: 0.04–1.2 mA/cm^2 V: 300 mL, t: 480 min	x_{PFOA}: 85%	F$^-$ deposition on the BDD surface.	[24]
Anode: BDD Cathode: Pt Area: 5.5 cm^2 Spacing: 20 mm	[PFOA]$_0$: 200 mg/L Electrolyte: 7.1 g/L Na$_2$SO$_4$ P: 0.3 MPa, T: 80–120 °C, j: 20 mA/cm^2 V: 400 mL, t: 360 min	x_{PFOA}: 95% x_{TOC}: 90% x_F^-: 90%	High temperature process greatly enhances the PFOA degradation in relation to the room temperature system.	[25]

2. Results and Discussion

In order to test the influence of the cathode material on the degradation process, a BDD anode was successively coupled with cathodes made of BDD, Pt, Zr, and stainless steel. Results of electrolysis runs conducted at 7.9 mA/cm^2, at 25 °C, for the treatment of 100 mg/L PFOA solutions (namely, 0.242 mol/m^3), are presented in Figure 1a. For each couple of electrodes, the curves show that PFOA concentration followed, from C$_{PFOA,0}$ = 100 mg/L to C$_{PFOA,t}$ ≈ 25 mg/L, a similar decrease, characteristic of a pseudo-first order kinetics. For experiments presented in Figure 1a the applied current density was higher than the limiting current density. Considering a pure mass transport controlled reaction for the first exchange of charge between a molecule of PFOA and the anode surface, the limiting current density calculated using the equation established from the Nernst diffusion model: $j_{lim} = n \cdot F \cdot k_m \cdot C_{PFOA,0}$ equals to 0.63 A/m^2 for n = 1, F = 96,485 C/mol, k_m = 2.7·10^{-5} m/s (determined experimentally using the ferri/ferro system, as described elsewhere [26]) for a flow rate of 0.360 m^3/h [27], C$_{PFOA,0}$ = 0.242 mol/m^3. This value of the limiting current density is more than 100 times lower than that applied during electrolysis (j = 79 A/m^2). Under these conditions, the decay of concentration from C$_{PFOA,0}$ to C$_{PFOA,t}$ depends upon the mass transfer coefficient k_m, the surface area (A) of the electrode, and the volume (V) of electrolyte [28], as follows:

$$C_{PFOA,t} = C_{PFOA,0} \cdot e^{\left(\frac{-t}{\tau}\right)} \qquad (2)$$

where the constant of time, τ, is defined by: τ = V/(k_m A). For its calculation we considered the following values V = 10^{-3} m^3, A = 63·10^{-4} m^2, and k_m = 2.7·10^{-5} m/s, obtaining a time constant (τ) equal to 5800 s. According to Equation (2), the PFOA theoretical concentration at t = 2 h is around 29 mg/L, which is in agreement with the experimental results (≈25 mg/L), as shown in Figure 1a.

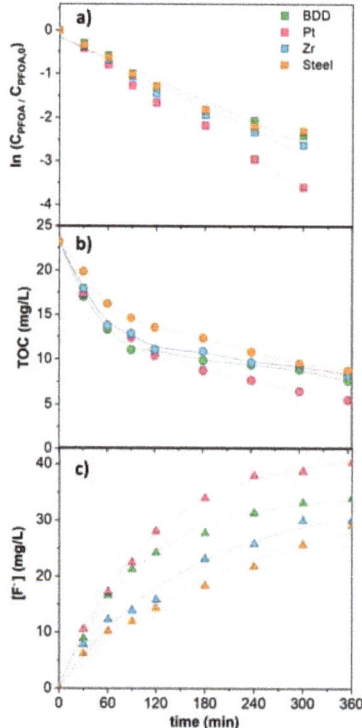

Figure 1. Influence of the cathode material in (**a**) PFOA removal (symbols: experimental data, lines: kinetic fitting), (**b**) TOC depletion, (**c**) released fluorine (symbols: experimental data, lines: kinetic fitting). Operating conditions: [PFOA]$_0$: 100 mg/L, j: 7.9 mA/cm^2, electrolyte: 3.5 mM Na$_2$SO$_4$, T: 25 °C, pH$_0$: 4.

The excess of charge on the first oxidation step of PFOA is used for other electron transfers and by the action of HO$^\bullet$ radicals during the degradation of the numerous intermediates. PFOA removal after 6 h was 100%, 98.1%, 97.9%, and 97.6% for Pt, steel, Zr, and BDD, respectively. Values of the rate constants and regression coefficients are collected in Table 2. It should be noted that the Pt cathode has the best results with a PFOA degradation rate 39% faster than the other tested materials, which exhibit a similar behavior between them. This enhancement is also reflected in the mineralization degree (Figure 1b), with a 76.1% Total Organic Carbon (TOC) removal with the BDD-Pt system.

Table 2. PFOA degradation and fluoride release kinetics.

	$k_{PFOA} \cdot 10^3$ (min^{-1})	r^2	$k_{F^-} \cdot 10^3$ (min^{-1})	r^2
BDD	8.92 ± 0.68	0.988	8.97 ± 0.28	0.999
Pt	11.86 ± 0.32	0.994	10.36 ± 0.44	0.997
Zr	8.16 ± 0.58	0.979	6.23 ± 0.86	0.982
Steel	7.98 ± 0.42	0.981	4.70 ± 0.78	0.982

As previously explained, one of the main challenges in PFOA oxidation is the effective breakdown of the C-F bond. PFOA defluorination was followed along the reaction by means of ionic chromatography, as depicted in Figure 1c. The trend for defluorination was Pt > BDD > Zr > steel. In this case, the cathode also played an important role, reaching a 58.6% in the case of Pt, against 42–49% for BDD,

Zr, and steel. Moreover, fluoride release follows a first order, as reflected in Figure 1 and Table 2, where the function of Pt as electrocatalyst is confirmed.

Pt is a common catalyst in hydrodehalogenation reaction of organic molecules, because of its capacity to adsorb hydrogen, providing a catalytic site were the dehalogenation takes place [29]. H_2 generation by water electrolysis on the cathode's surface may be responsible for PFOA hydrodefluorination, following the reaction mechanism shown in Figure 2.

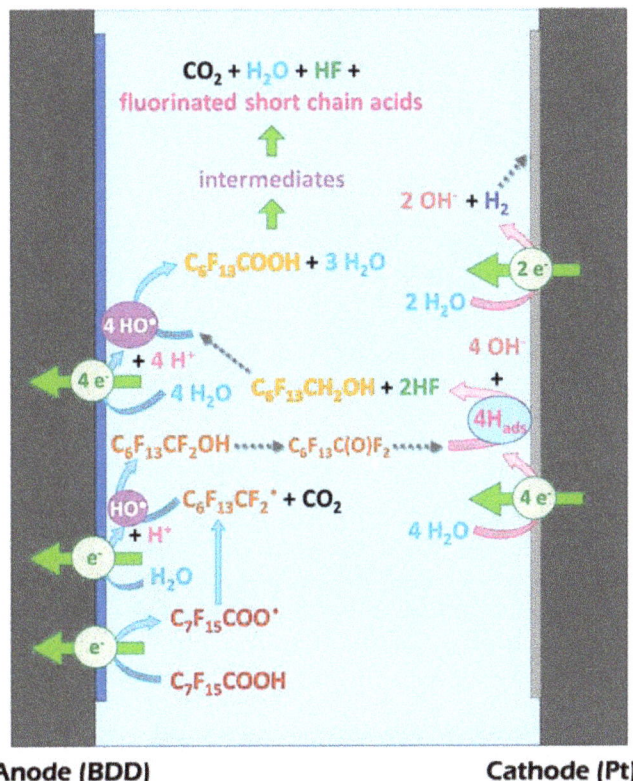

Figure 2. PFOA electrooxidation mechanism.

Because PFOA is inert to hydroxyl radicals, its degradation is initiated on the anode by a direct electron transfer reaction to form a perfluoro radical $C_7F_{15}COO^\bullet$ (Equation (3)). This radical loses its carboxylic group (Equation (4)) and reacts with $HO^\bullet$ leading to the generation of $C_7F_{13}{}^-CF_2OH$ (Equation (5)), as previously described by Zhang et al. [30].

$$BDD + C_7F_{15}COO^- \rightarrow BDD + C_7F_{15}COO^\bullet + e^- \tag{3}$$

$$C_7F_{15}COO^\bullet \rightarrow C_7F_{15}{}^\bullet + CO_2 \tag{4}$$

$$C_7F_{15}{}^\bullet + HO^\bullet \rightarrow C_7F_{15}OH \tag{5}$$

This alcohol then reacts according to three pathways, (for clarity reasons, only the first one (i) is illustrated in Figure 2):

(i) With adsorbed hydrogen generated by water electro-reduction at the cathode, releasing 2 F$^-$ (Equation (6)).

$$C_6F_{13}CF_2OH + 4H_{ads} \rightarrow C_6F_{13}CH_2OH + 2HF \qquad (6)$$

As the first carbon in the alkyl chain is now defluorinated, HO$^\bullet$ can attack it once again leading to the formation of $C_6F_{13}COOH$. This mechanism is similar to that presented for PFOA photocatalytic degradation by Wang et al. [20] and theoretic quantum calculations and experimental data collected by Trojanowicz et al. [31]. Hence, this step depends strongly on the cathode material.

(ii) With hydroxyl radicals leading to the formation of COF_2, as related by Niu et al. [31] and Zhang et al. [28], following Equations (7)–(9):

$$C_7F_{15}OH + HO^\bullet \rightarrow C_7F_{15}O^\bullet + H_2O \qquad (7)$$

$$C_7F_{15}O^\bullet \rightarrow C_6F_{13}^\bullet + COF_2 \qquad (8)$$

$$COF_2 + H_2O \rightarrow CO_2 + 2HF \qquad (9)$$

According to George et al. [30] hydrolysis of carbonyl fluoride COF_2 in the aqueous phase is extremely fast since its half-life is 0.7 s at T = 273 K.

(iii) Giving the perfluorocarbonyl fluoride (Equation (10)) for which hydrolysis leads the formation of perfluorocarboxylic acid and HF (Equation (11)) [30,32].

$$C_7F_{15}OH \rightarrow C_6F_{13}COF + HF \qquad (10)$$

$$C_6F_{13}COF + H_2O \rightarrow C_6F_{13}COO^- + HF + H^+ \qquad (11)$$

Considering this complex reaction mechanism, it should be noted that TOC decay was faster within the first hour of reaction, then it slowed down (Figure 1b). This is related to the generation of short-chain fluorinated acids (decarboxylation step), which are less active to electro-oxidation processes. In fact, pH value in the Pt system decreased from 4 to 3.2 in 120 min, maintaining this pH until the end of the reaction, which evidences the generation of these acidic species.

Data displayed in Figure 1 allow to determine the fluoride concentrations produced with respect to the degraded carbon in the form of CO_2 (F$^-$/CO_2) or with respect to the PFOA eliminated over time (F$^-$/PFOA). Figure 3 shows that the PFOA defluorination leads to the formation of 1 to 1.5 fluorine ions per removed atom of carbon in the first three hours of electrolysis. According to the proposed mechanism, this value close to the one at the beginning of the electrolysis is related to decarboxylation which leads to the formation of R_f-COF. The kinetics of this step are probably faster than that of the defluorination stages according to Equations (6)–(9). Besides, part of the process can be attributable to the electrocatalytic hydrogenation of the perfluorocarbonyl fluoride $C_6F_{13}COF$ that forms simultaneously the hydrofluoric acid and the 1,1-dihydroperfluoroalkyl alcohol $C_6F_{13}CH_2OH$. This alcohol is stable but easily oxidizable on the BDD anode [33] (cf. Figure 2). This process is slowed down by the diffusion of the species to the cathode. Not all molecules undergo the loss of two fluoride atoms, which would explain the value of 1.5 instead of the usual ratio 1.9 present in the initial PFOA molecule.

In addition, Figure 3 shows the variation of the ratio between the concentration of fluoride ions released and the concentration of the removed PFOA. This ratio varies from 7.7 to 9 for 360 min of electrolysis. These values highlight the high, yet incomplete PFOA defluorination. Finally, the ratio between the carbon loss (in the form of CO_2) and the removed PFOA (CO_2/PFOA) is in the order of 6–7, slightly less than 8, i.e., the theoretical value for $C_7F_{15}COOH$, confirming the formation of reaction intermediates. This ratio decreases during electrolysis: the degradation being faster at the beginning of the reaction, until t:150 min.

At this time more than 85% of PFOA has been eliminated. PFOA depletion slows down both the defluorination and carbon skeleton breaking. In addition, shorter molecular chains could display slower kinetics.

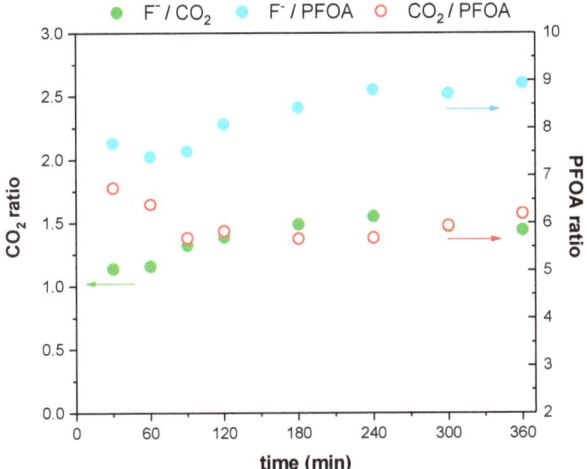

Figure 3. Variation of fluorine ions (full symbols) and carbon removal (empty symbols) during electrolysis with respect to carbon removal and degraded PFOA. Operating conditions: [PFOA]$_0$: 100 mg/L, j: 7.9 mA/cm^2, electrolyte: 3.5 mM Na$_2$SO$_4$, T: 25 °C, pH$_0$: 4.

Figure 4 shows the ratio of fluorine and carbon atoms contained in the chemical intermediates. The molar concentration of F and C atoms contained in the intermediates are defined, respectively, as follows:

$$C_{F,\text{intermediates}} = 15 \cdot (C_{PFOA,0} - C_{PFOA,t}) - C_{F^-,t} \tag{12}$$

$$C_{C,\text{intermediates}} = TOC_t - TOC_{PFOA,t} \tag{13}$$

where $C_{PFOA,0}$ and $C_{PFOA,t}$ refer to the molar concentration of PFOA at initial time and at time t, respectively; $C_{F^-,t}$ is the molar concentration of fluorine ions at t; TOC_t and $TOC_{PFOA,t}$ are the total carbon molar concentration and the carbon molar concentration in the PFOA, respectively.

From Figure 1, after 360 min of electrolysis, the defluorination rate is 59% whereas more than 98% of PFOA and 76% of TOC have been eliminated. Figure 4 highlights that in this moment, the intermediates still contain 24% of carbon and 41% of fluorine. Xiao et al. reached a 90% defluorination and mineralization working at high temperature (T: 80–120 °C), meaning they managed to degrade the short-chain acids [25]. This is in agreement with the results for degradation of phenol in heterogeneous Fenton at high temperature, where maleic, malonic, oxalic, and formic acids can be completely degraded [11], in contrast with room temperature processes [34]. Aiming to verify Xiao et al.'s results, an electrooxidation run at 80 °C was performed using Pt cathode. After 30 min reaction there was an overvoltage on the cell due to the damage on the cathode, probably because of the HF attack (Figure S1 of the Supplementary Material). Hence, high temperature electrooxidation could not be performed in our system and further runs were conducted at 25 °C.

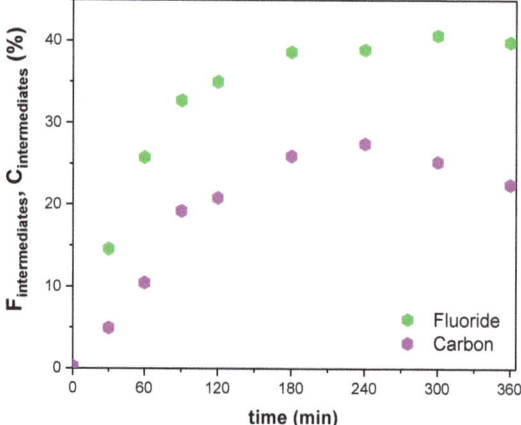

Figure 4. Ratio of fluorine and carbon atoms in the chemical intermediates. Operating conditions: [PFOA]$_0$: 100 mg/L, j: 7.9 mA/cm^2, electrolyte: 3.5 mM Na$_2$SO$_4$, T: 25 °C, pH$_0$: 4.

After selecting Pt as the best cathode, within the tested materials, different salts were used as electrolyte, NaClO$_4$, KNO$_3$, Na$_2$SO$_4$, Na$_2$S$_2$O$_8$, at 3.5 mM. Results for these experiments can be found in Figure 5. As previously reported by Schaefer et al. [21], the influence of the electrolyte type on PFOA degradation is very low. Still, significant differences were found for TOC removal, where the removal efficiency followed this trend Na$_2$SO$_4$ (76.1%) > Na$_2$S$_2$O$_8$ (72.6%) > KNO$_3$ (70.5%) > NaClO$_4$ (67.1%). Sulfate achieved both a slightly higher mineralization degree and defluorination. This can be explained by the fact that sulfate anions behave as an active electrolyte via the electrochemical generation of the strong oxidizing sulfate radicals (SO$_4$$^{\bullet-}$) on a BDD anode [35,36]. Indeed, the oxidation of water at the anode greatly decreases locally the pH at the surface leading to the formation of HSO$_4^-$ from SO$_4$$^{2-}$. Then HSO$_4^-$ reacts with HO$^\bullet$ radicals to form sulfate radicals [33,37].

$$HSO_4^- + HO^\bullet \rightarrow SO_4^{\bullet-} + H_2O \quad k = 6.9 \cdot 10^5 M^{-1} s^{-1} \tag{14}$$

SO$_4$$^{\bullet-}$ radical participates in electron transfer reactions and promotes the decarboxylation of carboxylic acids, contrary to HO$^\bullet$ which rather acts in hydrogen abstraction or addition [38]. In addition, sulfate radicals are more stable than hydroxyl radicals (their half-life is 30–40 µs and 10^{-3} µs, respectively).

Considering PFOA degradation with sulfate radicals, the literature review by Yang et al. highlights that the decomposition and defluorination efficiencies increase with a decrease in PFOA chain-length [39]. Besides the major role of hydroxyl radicals on PFOA oxidation, the presence of sulfate radicals helps to improve the degradation of the generated intermediates. Qian et al. [40] estimated the constant rate of PFOA degradation with sulfate radicals at 2.59·10^5 M^{-1}s^{-1}. This is consistent with the higher TOC removal observed in our experiments in presence of sulfate. Furthermore, sulfate is a more environmentally friendly electrolyte, in comparison to perchlorate and nitrate, which can be considered pollutants by themselves. Thus, the rest of experiments were carried out using Na$_2$SO$_4$ 3.5 mM.

Influence of initial pH (pH$_0$) on PFOA degradation was also evaluated working at the natural pH of PFOA solution (pH: 4) and at pH values of 7 and 9. Results for these experiments are shown in Figure 6. As it may be seen in Figure 6d, reaction media is quickly acidified. This is related to both the generation of short chain acids and the reaction between sulfate radicals and water to produce hydroxyl radicals, which also generates protons, as depicted in Equation (15). PFOA decay (Figure 6a) was similar for all the runs. However, pH$_0$ had a great influence on the initial rate for TOC abatement, related to the higher oxidation potential of sulfate radicals in alkaline media [41]. Despite achieving a higher mineralization degree at pH$_0$: 9, the highest defluorination was reached when starting in acidic media.

$$SO_4^{\bullet-} + H_2O \rightarrow H^+ + HO^\bullet + SO_4^{2-} \qquad (15)$$

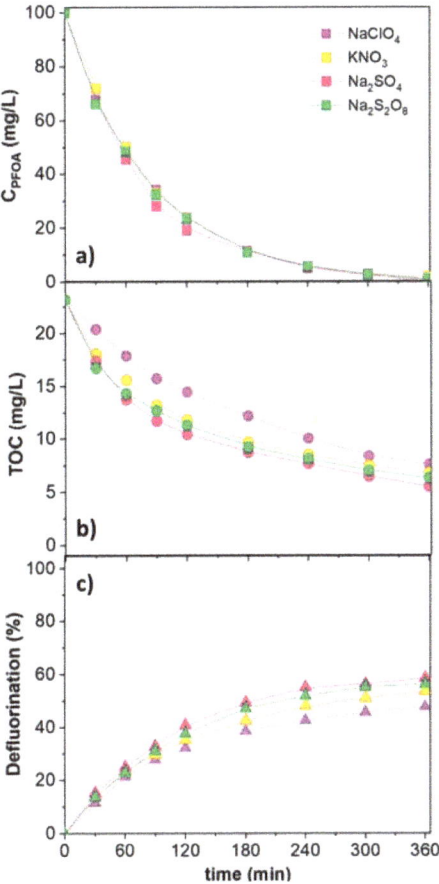

Figure 5. Influence of the electrolyte in PFOA (**a**) degradation, (**b**) mineralization, and (**c**) defluorination in electrooxidation using BDD/Pt electrodes. Operating conditions: [PFOA]$_0$: 100 mg/L, j: 7.9 mA/cm^2, electrolyte: 3.5 mM, T: 25 °C, pH$_0$: 4.

So far, the role of the cathode as electrocatalyst in the degradation and defluorination of PFOA has been proved. Also, the influence of several operating conditions has been tested, demonstrating an overall great decontamination working at low electrolyte concentration at mild temperature. Nonetheless, in order to compare the obtained results with those reported in literature, we have compared the defluorination degree against the energetic requirements, measured as the applied charge, as shown in Figure 7. As may be seen, both Shaefer et al. [21] and Urtiaga et al. [22] boosted the defluorination degree when increasing the applied charge. However, the results presented in this work using Pt cathode at 7.9 mA/cm^2, 3.5 mM Na$_2$SO$_4$ at pH$_0$:4 and T:25 °C are the most competitive in terms of PFOA defluorination against electric charge. In this sense, cathode selection becomes a key point for both increasing the activity and reducing the energy requirements in PFOA electrooxidation.

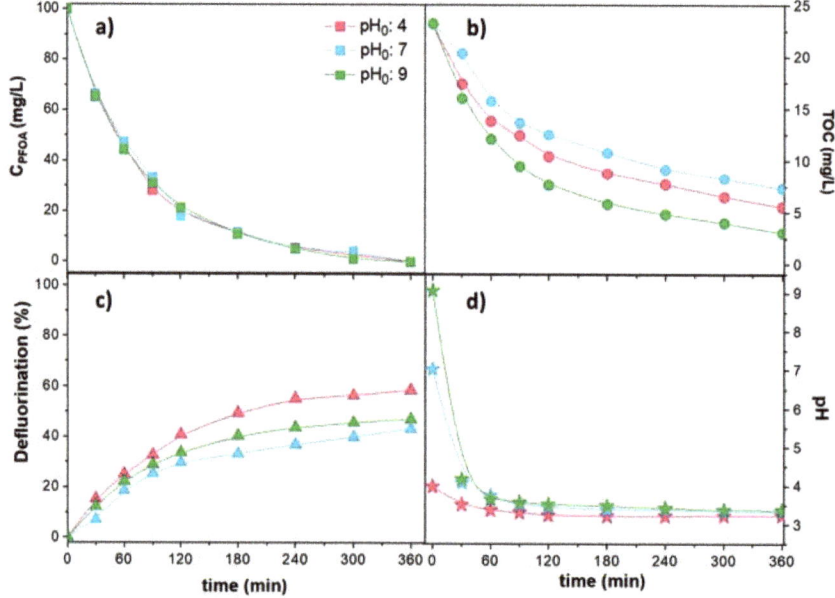

Figure 6. Influence of the pH_0 in PFOA (**a**) degradation, (**b**) mineralization, and (**c**) defluorination and (**d**) pH evolution in electrooxidation using BDD/Pt electrodes. Operating conditions: [PFOA]$_0$: 100 mg/L, j: 7.9 mA/cm^2, Na$_2$SO$_4$: 3.5 mM, T: 25 °C.

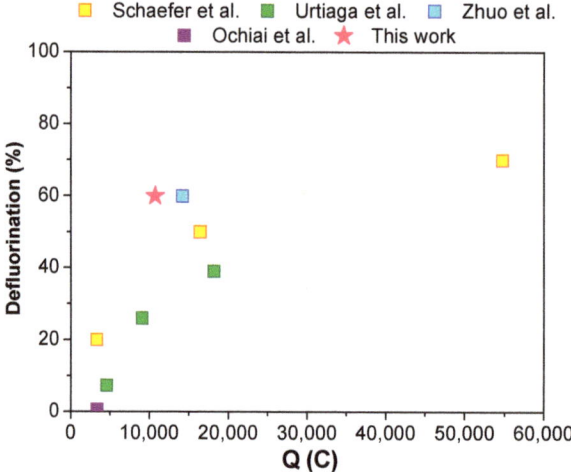

Figure 7. Process comparison in terms of defluorination against electric charge for PFOA electrooxidation with BDD anodes. Cathodes: Schaefer et al.—W [21], Urtiaga et al.—W [22], Zhuo et al.—BDD [23], Ochiai et al.—Pt [24], this work—Pt.

3. Materials and Methods

3.1. Reactants

Perfluoroctanoic acid (95 wt.%), Na_2SO_4, KNO_3, $NaClO_4$, $Na_2S_2O_8$, acetonitrile (ACN), H_2SO_4, and NaOH were supplied by Sigma-Aldrich (Darmstadt, Germany). All reagents are of analytical grade and they were used as received without further purification. Working standard solutions of PFOA and fluoride (NaF from Sigma-Aldrich, Darmstadt, Germany) were prepared for calibration.

3.2. Experimental Set-Up

The electrochemical oxidation system consists in a 1-L thermoregulated glass reservoir connected to the cell through a centrifugal pump. PFOA solution was recycled in the system at 360 L/h flow rate and the temperature was set at 25 ± 1 °C. The electrochemical cell is a one-compartment flow filter-press reactor which was operated under galvanostatic conditions using an ELCAL 924 power supply (Italy). Electrodes present a 63 cm^2 active surface and the gap between them was set at 10 mm. A detailed scheme of the experimental set-up can be found elsewhere [42]. All experiences were performed with a BDD anode from Adamant Technologies (La Chaux-de-Fonds, Switzerland), which was elaborated by chemical vapor deposition on a conductive substrate of Si. BDD (Adamant Technologies, La Chaux-de-Fonds, Switzerland), Zirconium, Stainless steel, and Pt (5 µm) on titanium substrate (provided by MAGNETO special anodes B.V., Schiedam, Netherlands) were employed as cathodes. Before each electrolysis, the working electrodes were anodically pretreated (40 mA/cm^2 for 30 min in 0.1 M H_2SO_4) to clean their surfaces of any possible adsorbed impurities. Then, the system was rinsed by ultrapure water.

In a typical reaction 1 L PFOA solution (100 mg/L) with 3.5 mM Na_2SO_4 as electrolyte at the natural pH of the solution (pH:4) was loaded to the reservoir, preheated to 25 °C and recycled through the system. Once the selected temperature was reached, the power supply was turned on and current intensity was set at 0.5 A, representing this as the reaction starting time. Samples were taken at regular intervals in the tank. The global volume of samples was less than 10% of the total volume. All runs were performed by triplicate with a deviation lower than 5% in all cases.

3.3. Analytical Methods

Samples were periodically withdrawn from the reactors, filtered through 0.2 µm nylon syringe plug-in filters and immediately analyzed, without any further manipulation. PFOA concentration was measured by high performance liquid chromatography connected with an ultraviolet-visible spectrometry detector (HPLC-UV Agilent 1200 Series HPLC, Santa Clara, USA). An ion-exclusion column (ZORBAX Eclipse Plus C18, 100 mm, 1.8 µm, Agilent, USA) was used as the stationary phase. As mobile phase mixture of ACN/4 mM H_2SO_4 aqueous solution with a ratio: 3/2 was employed and the column temperature was set to 50 °C. A 60% CAN—40% mixture was employed at 0.5 mL/min. The detection UV wavelength was set to 206 nm. Total organic carbon was quantified using a TOC analyzer (Shimadzu TOC-VSCH, Kyoto, Japan). Fluoride was analyzed in an ion chromatograph with chemical suppression (Metrohm 790 IC, Herisau, Switzerland) using a conductivity detector. A Metrosep A supp 5–250 column (25 cm long, 4 mm diameter, Herisau, Switzerland) was used as the stationary phase and 0.7 mL/min of a 3.2 mM/1 mM aqueous solution of Na_2CO_3 and $NaHCO_3$, respectively, as the mobile phase.

4. Conclusions

PFOA electro-degradation follows a complex mechanism which involves both oxidation reactions on the anode surface and reduction reactions, responsible for the molecule's defluorination, which take place over the cathode. Electrocatalytic hydrogenation of the unsaturated acyl fluoride R_fCOF can be a route for the degradation process. Atomic hydrogen produced in situ at the catalyst surface can form simultaneously the alcohol $RFCH_2OH$ and hydrofluoric acid.

In this work, different cathodes have been used, finding that its selection plays a key role in PFOA degradation. In this sense, Pt acts as an electrocatalyst because of its higher capacity to produce in situ atomic hydrogen, which seems efficient in hydrodefluorination. It has been also demonstrated that working at low electrolyte concentration (3.5 mM Na_2SO_4), complete PFOA removal can be reached with up to 76.1% TOC abatement and 58.6% defluorination working at the natural pH of the solution (pH_0: 4). The kind of electrolyte employed did not have a significant impact on the overall reaction. Still, slightly better results were achieved using sulfate because of the generation of sulfate radicals. Regarding the influence of the starting pH, higher TOC removal was obtained working at pH_0: 9, while at higher pH values PFOA mineralization was hindered. When comparing the results obtained in this work with those reported in literature, it must be remarked that the employed BDD-Pt system allows a higher defluorination degree with a lower energy consumption. In view to render the process economically viable to treat dilute solutions, further experiments are planned to combine the electrochemical process with a preconcentration step (such as filtration or adsorption).

Supplementary Materials: The following are available online at http://www.mdpi.com/2073-4344/10/8/902/s1, Figure S1. Damaged Pt cathode after high temperature PFOA electrooxidation (T: 80 °C).

Author Contributions: A.L.G.C.: conceptualization, investigation, data curation, methodology, writing: original draft. J.A.Z.: supervision, writing: review. A.S.: supervision, data curation, validation, writing: review. K.G.S.: formal analysis, supervision, validation, writing: review. J.A.C.: formal analysis, funding acquisition, supervision, validation, writing: review. All authors have read and agreed to the published version of the manuscript.

Funding: This research was funded by the Spanish Ministerio de Ciencia, Innovación y Universidades through project CTM2016-76454-R and by Comunidad de Madrid by P2018/EMT-4341 REMTAVARES-CM.

Acknowledgments: Authors thank the funding received from Ministerio de Ciencia, Innovación y Universidades through research project CTM2016-76454-R and Comunidad de Madrid for P2018/EMT-4341 REMTAVARES-CM. Alicia L. Garcia-Costa would like to thank Campus France for the mobility grant under the Make Our Planet Great Again (MOPGA) program and the Spanish Ministerio de Ciencia, Innovación y Universidades for mobility grant EST2019-013106-I. She would also like to thank both the Spanish Ministerio de Economía y Competitividad and the European Social Fund for the PhD grant BES-2014-067598.

Conflicts of Interest: The authors declare no conflict of interest.

References

1. Gebbink, W.A.; van Leeuwen, S.P.J. Environmental Contamination and Human Exposure to PFASs Near a Fluorochemical Production Plant: Review of Historic and Current PFOA and GenX Contamination in the Netherlands. *Environ. Int.* **2020**, *137*, 11. [CrossRef] [PubMed]
2. Ao, J.J.; Yuan, T.; Xia, H.; Ma, Y.N.; Shen, Z.M.; Shi, R.; Tian, Y.; Zhang, J.; Ding, W.J.; Gao, L.; et al. Characteristic and Human Exposure Risk Assessment of Per- and Polyfluoroalkyl Substances: A study Based on Indoor Dust and Drinking Water in China. *Environ. Pollut.* **2019**, *254*, 9. [CrossRef] [PubMed]
3. Hori, H.; Hayakawa, E.; Einaga, H.; Kutsuna, S.; Koike, K.; Ibusuki, T.; Kiatagawa, H.; Arakawa, R. Decomposition of Environmentally Persistent Perfluorooctanoic Acid in Water by Photochemical Approaches. *Environ. Sci. Technol.* **2004**, *38*, 6118–6124. [CrossRef] [PubMed]
4. Kannan, K.; Koistinen, J.; Beckmen, K.; Evans, T.; Gorzelany, J.F.; Hansen, K.J.; Jones, P.D.; Helle, E.; Nyman, M.; Giesy, J.P. Accumulation of Perfluorooctane Sulfonate in Marine Mammals. *Environ. Sci. Technol.* **2001**, *35*, 1593–1598. [CrossRef]
5. Wang, G.H.; Wang, X.L.; Xing, Z.N.; Lu, J.J.; Chang, Q.G.; Tong, Y.B. Occurrence and Distribution of Perfluorooctane Sulfonate and Perfluorooctanoic Acid in Three Major Rivers of Xinjiang, China. *Environ. Sci. Pollut. Res.* **2019**, *26*, 28062–28070. [CrossRef]
6. Pico, Y.; Blasco, C.; Farre, M.; Barcelo, D. Occurrence of Perfluorinated Compounds in Water and Sediment of L'Albufera Natural Park (Valencia, Spain). *Environ. Sci. Pollut. Res.* **2012**, *19*, 946–957. [CrossRef]
7. Domingo, J.L.; Nadal, M. Human Exposure to Per-And Polyfluoroalkyl Substances (PFAS) through Drinking Water: A Review of the Recent Scientific Literature. *Environ. Res.* **2019**, *177*, 10. [CrossRef]
8. Saeidi, N.; Kopinke, F.D.; Georgi, A. Understanding the Effect of Carbon Surface Chemistry on Adsorption of Perfluorinated Alkyl Substances. *Chem. Eng. J.* **2020**, *381*, 11. [CrossRef]

9. Wang, F.; Shih, K.M. Adsorption of Perfluorooctanesulfonate (PFOS) and Perfluorooctanoate (PFOA) on Alumina: Influence of Solution pH and Cations. *Water Res.* **2011**, *45*, 2925–2930.
10. Maimaiti, A.; Deng, S.B.; Meng, P.P.; Wang, W.; Wang, B.; Huang, J.; Wang, Y.J.; Yu, G. Competitive Adsorption of Perfluoroalkyl Substances on Anion Exchange Resins in Simulated AFFF-Impacted Groundwater. *Chem. Eng. J.* **2018**, *348*, 494–502. [CrossRef]
11. Garcia-Costa, A.L.; Zazo, J.A.; Casas, J.A. Microwave-Assisted Catalytic Wet Peroxide Oxidation: Energy Optimization. *Sep. Purif. Technol.* **2019**, *215*, 62–69. [CrossRef]
12. Garcia-Costa, A.L.; Zazo, J.A.; Rodriguez, J.J.; Casas, J.A. Intensification of Catalytic Wet Peroxide Oxidation With Microwave Radiation: Activity and Stability of Carbon Materials. *Sep. Purif. Technol.* **2019**, *209*, 301–306. [CrossRef]
13. Maruthamuthu, P.; Padmaja, S.; Huie, R.E. Rate Constants for Some Reactions of Free-Radicals with Haloacetates in Aqueous-Solution. *Int. J. Chem. Kinet.* **1995**, *27*, 605–612. [CrossRef]
14. Moriwaki, H.; Takagi, Y.; Tanaka, M.; Tsuruho, K.; Okitsu, K.; Maeda, Y. Sonochemical Decomposition of Perfluorooctane Sulfonate and Perfluorooctanoic Acid. *Environ. Sci. Technol.* **2005**, *39*, 3388–3392. [CrossRef] [PubMed]
15. Javed, H.; Lyu, C.; Sun, R.N.; Zhang, D.N.; Alvarez, P.J.J. Discerning the Inefficacy of Hydroxyl Radicals during Perfluorooctanoic Acid Degradation. *Chemosphere* **2020**, *247*, 6. [CrossRef]
16. Santos, A.; Rodriguez, S.; Pardo, F.; Romero, A. Use of Fenton Reagent Combined with Humic Acids for the Removal of PFOA from Contaminated Water. *Sci. Total Environ.* **2016**, *563*, 657–663. [CrossRef]
17. Chen, M.J.; Lo, S.L.; Lee, Y.C.; Huang, C.C. Photocatalytic Decomposition of Perfluorooctanoic Acid by Transition-Metal Modified Titanium Dioxide. *J. Hazard. Mater.* **2015**, *288*, 168–175. [CrossRef]
18. Chen, M.J.; Lo, S.L.; Lee, Y.C.; Kuo, J.; Wu, C.H. Decomposition of Perfluorooctanoic Acid by Ultraviolet Light Irradiation with Pb-Modified Titanium Dioxide. *J. Hazard. Mater.* **2016**, *303*, 111–118. [CrossRef]
19. Gomez-Ruiz, B.; Ribao, P.; Diban, N.; Rivero, M.J.; Ortiz, I.; Urtiaga, A. Photocatalytic Degradation and Mineralization of Perfluorooctanoic Acid (PFOA) Using a Composite TiO_2–rGO Catalyst. *J. Hazard. Mater.* **2018**, *344*, 950–957. [CrossRef]
20. Wang, S.N.; Yang, Q.; Chen, F.; Sun, J.; Luo, K.; Yao, F.B.; Wang, X.L.; Wang, D.B.; Li, X.M.; Zeng, G.M. Photocatalytic Degradation of Perfluorooctanoic Acid and Perfluorooctane Sulfonate in Water: A Critical Review. *Chem. Eng. J.* **2017**, *328*, 927–942. [CrossRef]
21. Schaefer, C.E.; Andaya, C.; Burant, A.; Condee, C.W.; Urtiaga, A.; Strathmann, T.J.; Higgins, C.P. Electrochemical Treatment of Perfluorooctanoic Acid and Perfluorooctane Sulfonate: Insights into Mechanisms and Application to Groundwater Treatment. *Chem. Eng. J.* **2017**, *317*, 424–432. [CrossRef]
22. Urtiaga, A.; Fernandez-Gonzalez, C.; Gomez-Lavin, S.; Ortiz, I. Kinetics of the Electrochemical Mineralization of Perfluorooctanoic Acid on Ultrananocrystalline Boron Doped Conductive Diamond Electrodes. *Chemosphere* **2015**, *129*, 20–26. [CrossRef] [PubMed]
23. Zhuo, Q.F.; Deng, S.B.; Yang, B.; Huang, J.; Wang, B.; Zhang, T.T.; Yu, G. Degradation of Perfluorinated Compounds on a Boron-Doped Diamond Electrode. *Electrochim. Acta* **2012**, *77*, 17–22. [CrossRef]
24. Ochiai, T.; Iizuka, Y.; Nakata, K.; Murakami, T.; Tryk, D.A.; Fujishima, A.; Koide, Y.; Morito, Y. Efficient Electrochemical Decomposition of Perfluorocarboxylic Acids by the Use of a Boron-Doped Diamond Electrode. *Diam. Relat. Mater.* **2011**, *20*, 64–67. [CrossRef]
25. Xiao, H.S.; Lv, B.Y.; Zhao, G.H.; Wang, Y.J.; Li, M.F.; Li, D.M. Hydrothermally Enhanced Electrochemical Oxidation of High Concentration Refractory Perfluorooctanoic Acid. *J. Phys. Chem. A* **2011**, *115*, 13836–13841. [CrossRef]
26. Mais, L.; Mascia, M.; Palmas, S.; Vacca, A. Photoelectrochemical Oxidation of Phenol With Nanostructured TiO_2-PANI Electrodes under Solar Light Irradiation. *Sep. Purif. Technol.* **2019**, *208*, 153–159. [CrossRef]
27. Weiss, E.; Groenen-Serrano, K.; Savall, A.; Comninellis, C. A Kinetic Study of the Electrochemical Oxidation of Maleic Acid on Boron Doped Diamond. *J. Appl. Electrochem.* **2007**, *37*, 41–47. [CrossRef]
28. Walsh, F.C. A First Course in Electrochemical Engineering. *The Electrochemical Consultancy.* **1993**, 381. [CrossRef]
29. Pizarro, A.H.; Molina, C.B.; Fierro, J.L.G.; Rodriguez, J.J. On the Effect of Ce Incorporation on Pillared Clay-Supported Pt and Ir Catalysts for Aqueous-Phase Hydrodechlorination. *Appl. Catal. B Environ.* **2016**, *197*, 236–243. [CrossRef]

30. Zhang, Y.Y.; Moores, A.; Liu, J.X.; Ghoshal, S. New Insights into the Degradation Mechanism of Perfluorooctanoic Acid by Persulfate from Density Functional Theory and Experimental Data. *Environ. Sci. Technol.* **2019**, *53*, 8672–8681. [CrossRef]
31. Trojanowicz, M.; Bojanowska-Czajka, A.; Bartosiewicz, I.; Kulisa, K. Advanced Oxidation/Reduction Processes Treatment for Aqueous Perfluorooctanoate (PFOA) and Perfluorooctanesulfonate (PFOS)—A Review of Recent Advances. *Chem. Eng. J.* **2018**, *336*, 170–199. [CrossRef]
32. Niu, J.F.; Lin, H.; Gong, C.; Sun, X.M. Theoretical and Experimental Insights into the Electrochemical Mineralization Mechanism of Perfluorooctanoic Acid. *Environ. Sci. Technol.* **2013**, *47*, 14341–14349. [CrossRef] [PubMed]
33. Stefanova, A.; Ayata, S.; Erem, A.; Ernst, S.; Baltruschat, H. Mechanistic Studies on Boron-Doped Diamond: Oxidation of Small Organic Molecules. *Electrochim. Acta* **2013**, *110*, 560–569. [CrossRef]
34. Zazo, J.A.; Casas, J.A.; Mohedano, A.F.; Rodriguez, J.J. Catalytic Wet Peroxide Oxidation of Phenol with a Fe/active Carbon Catalyst. *Appl. Catal. B Environ.* **2006**, *65*, 261–268. [CrossRef]
35. Serrano, K.; Michaud, P.A.; Comninellis, C.; Savall, A. Electrochemical Preparation of Peroxodisulfuric Acid Using Boron Doped Diamond Thin Film Electrodes. *Electrochim. Acta* **2002**, *48*, 431–436. [CrossRef]
36. Lan, Y.D.; Coetsier, C.; Causserand, C.; Serrano, K.G. On the Role of Salts for the Treatment of Wastewaters Containing Pharmaceuticals by Electrochemical Oxidation Using a Boron Doped Diamond Anode. *Electrochim. Acta* **2017**, *231*, 309–318. [CrossRef]
37. Kolthoff, I.M.; Miller, I.K. The Chemistry of Persulfate.1. The Kinetics and Mechanism of the Decomposition of the Persulfate ion in Aqueous Medium. *J. Am. Chem. Soc.* **1951**, *73*, 3055–3059. [CrossRef]
38. Neta, P.; Madhavan, V.; Zemel, H.; Fessenden, R.W. Rate Constants and Mechanism of Reaction of Sulfate Radical Anion with Aromatic Compounds. *J. Am. Chem. Soc.* **1977**, *99*, 163–164. [CrossRef]
39. Yang, L.; He, L.Y.; Xue, J.M.; Ma, Y.F.; Xie, Z.Y.; Wu, L.; Huang, M.; Zhang, Z.L. Persulfate-Based Degradation of Perfluorooctanoic Acid (PFOA) and Perfluorooctane Sulfonate (PFOS) in Aqueous Solution: Review on Influences, Mechanisms and Prospective. *J. Hazard. Mater.* **2020**, *393*, 11. [CrossRef]
40. Qian, Y.J.; Guo, X.; Zhang, Y.L.; Peng, Y.; Sun, P.Z.; Huang, C.H.; Niu, J.F.; Zhou, X.F.; Crittenden, J.C. Perfluorooctanoic Acid Degradation Using UV-Persulfate Process: Modeling of the Degradation and Chlorate Formation. *Environ. Sci. Technol.* **2016**, *50*, 772–781. [CrossRef]
41. Silveira, J.E.; Garcia-Costa, A.L.; Cardoso, T.O.; Zazo, J.A.; Casas, J.A. Indirect Decolorization of Azo Dye Disperse Blue 3 by Electro-Activated Persulfate. *Electrochim. Acta* **2017**, *258*, 927–932. [CrossRef]
42. Lan, Y.; Coetsier, C.; Causserand, C.; Serrano, K.G. An Experimental and Modelling Study of the Electrochemical Oxidation Of Pharmaceuticals Using a Boron-Doped Diamond Anode. *Chem. Eng. J.* **2018**, *333*, 486–494. [CrossRef]

 © 2020 by the authors. Licensee MDPI, Basel, Switzerland. This article is an open access article distributed under the terms and conditions of the Creative Commons Attribution (CC BY) license (http://creativecommons.org/licenses/by/4.0/).

Article

Simulated Ageing of Crude Oil and Advanced Oxidation Processes for Water Remediation since Crude Oil Pollution

Filomena Lelario [1], Giuliana Bianco [1], Sabino Aurelio Bufo [1,2,*] and Laura Scrano [3]

[1] Department of Sciences, University of Basilicata, Via dell'Ateneo Lucano 10, 85100 Potenza, Italy; filomena.lelario@unibas.it (F.L.); giuliana.bianco@unibas.it (G.B.)
[2] Department of Geography, Environmental Management & Energy Studies, University of Johannesburg, Johannesburg 2092, South Africa
[3] Department of European Cultures (DICEM), University of Basilicata, 75100 Matera, Italy; laura.scrano@unibas.it
* Correspondence: sabino.bufo@unibas.it; Tel.: +39-0971-6237

Abstract: Crude oil can undergo biotic and abiotic transformation processes in the environment. This article deals with the fate of an Italian crude oil under simulated solar irradiation to understand (i) the modification induced on its composition by artificial ageing and (ii) the transformations arising from different advanced oxidation processes (AOPs) applied as oil-polluted water remediation methods. The AOPs adopted were photocatalysis, sonolysis and, simultaneously, photocatalysis and sonolysis (sonophotocatalysis). Crude oil and its water-soluble fractions underwent analysis using GC-MS, liquid-state ^{1}H-NMR, Fourier transform ion cyclotron resonance mass spectrometry (FT-ICR-MS), and fluorescence. The crude oil after light irradiation showed (i) significant modifications induced by the artificial ageing on its composition and (ii) the formation of potentially toxic substances. The treatment produced oil oxidation with a particular effect of double bonds oxygenation. Non-polar compounds present in the water-soluble oil fraction showed a strong presence of branched alkanes and a good amount of linear and aromatic alkanes. All remediation methods utilised generated an increase of C_5 class and a decrease of C_6-C_9 types of compounds. The analysis of polar molecules elucidated that oxygenated compounds underwent a slight reduction after photocatalysis and a sharp decline after sonophotocatalytic degradation. Significant modifications did not occur by sonolysis.

Keywords: crude oil; photocatalysis; sonolysis; sonophotocatalysis; FT-ICR/MS; Kendrick plot; van Krevelen diagram; water; pollution; remediation

Citation: Lelario, F.; Bianco, G.; Bufo, S.A.; Scrano, L. Simulated Ageing of Crude Oil and Advanced Oxidation Processes for Water Remediation since Crude Oil Pollution. *Catalysts* 2021, 11, 954. https://doi.org/10.3390/catal11080954

Academic Editor: Fernando J. Beltrán Novillo

Received: 13 July 2021
Accepted: 4 August 2021
Published: 10 August 2021

Publisher's Note: MDPI stays neutral with regard to jurisdictional claims in published maps and institutional affiliations.

Copyright: © 2021 by the authors. Licensee MDPI, Basel, Switzerland. This article is an open access article distributed under the terms and conditions of the Creative Commons Attribution (CC BY) license (https://creativecommons.org/licenses/by/4.0/).

1. Introduction

The composition of petroleum crude oil varies widely depending on the source and processing. Oil is a complex organic mixture counting for a high number of chemically distinct components, including unsaturated and saturated hydrocarbons, hetero-atoms (such as N, S, and O) and a minor percentage of metals predominantly vanadium, nickel, iron, and copper. Many oil constituents can be carcinogens, neurotoxins, respiratory irritants, hepatotoxins, nephrotoxins, and mutagens. Their toxic effects can be acute and chronic, causing many direct symptoms and major long-term injuries, including reproductive problems and cancer [1].

The hydrocarbon fraction can be as high as 90% by weight in light oils, compared to about 70% in heavy crude oil. A majority of the heteroatomic free constituents are side-by-side paraffinic chains, naphthalene rings, and aromatic rings. Heteroatomic compounds constitute a relatively small portion of crude oils, less than 15%. However, they have significant implications since their presence, composition, and solubility, which depend on the origin of the crude oil, can cause either positive or negative effects in the transformation processes and are of environmental concern [2,3].

A significant consideration of the several processes affecting the crude oil spilt into the environment is needed to clarify the effects of increasingly widespread harmful events and predict the future fate of the oil. For this reason, the awareness of such phenomena will prove to be a valuable resource in the effort to develop innovative remediation technologies. The ecological impact of oil contamination in different environmental sections (marine, terrestrial and atmospheric) is a source of severe concern. Extraction techniques, transportation and refinery treatments of crude oil can originate pollution phenomena due to the dispersion of these compounds everywhere. These problems have attracted significant attention to understanding the fate of oil in the environment and the natural mechanisms of oil degradation and transformation to suggest a method to reduce the damages caused by original and derivative products [4–6].

As all xenobiotic substances, crude oil undergoes biotic (biotransformation by aquatic organisms such as algae, bacteria) and abiotic (hydrolysis, oxidation, photodegradation) processes, giving rise to many derivatives. In the same way as their parent molecules, these transformation products can lead to the contamination of terrestrial and aquatic environments due to oil deposition on soil and into the surface- and ground-water. Nevertheless, they can be more persistent and toxic than the parent compounds [1–3].

Extensive literature is already available on the microbiological degradation of crude oil, which received considerable attention from researchers. For example, since 1975, the biodegradability of crude oils has been studied and found to be highly dependent on their composition and incubation temperature [5]. Researchers also examined the ability of microorganisms to degrade a high number of hydrocarbons of a different structure in petroleum [6]. Furthermore, many authors have elucidated that the lighter fractions can undergo degradation more rapidly than the heavier ones, e.g., n-alkanes degraded more quickly than branched alkanes, and aromatics with two to three rings readily biodegraded through several pathways [4–8].

Photochemical processes are also essential contributors to pollutants' degradation and the removal of exogenous substances from the environment [9,10], especially in tropical and sub-tropical climates. In those areas, solar irradiation intensity is high, and the lack of nutrients hinders biological processes. Moreover, photochemical reactions are the primary cause of the compositional change of crude oil spilt in a marine environment [11–13]. Photolysis plays an essential role in the mousse formation that begins a few moments after an oil spill [12]. Due to sunlight, the interfacial tension of a crude oil film rapidly decreases, and chocolate mousse starts to form, which leads to the stabilisation of the water-in-oil emulsions [13,14]. The formation of emulsions seems to depend on the amount of asphaltene present in the oil film, and researchers reported that this amount increases upon irradiation [13]. Moreover, an increase in emulsion viscosity occurs due to the structural organisation of the asphaltenes [14].

The oxidised products resulting from the photochemical transformation significantly affect the viscosity, mousse formation, and weathered petroleum's physical properties. Moreover, photo-oxidation can lead to the destruction of existing toxic components, the generation of new toxic constituents and the formation of water-soluble products [10–14].

Since crude oil settles on the surface of water and soil, it undergoes solar irradiation. Solar degradation is a natural way for petroleum decontamination, also suggesting that techniques based on light irradiation could be helpful to the petroleum degradation processes. Light irradiation-based technologies have been improved using catalysts, the most effective and cheapest water purification tool being titanium dioxide (TiO_2) [15,16]. Researchers have exploited combinations of different advanced oxidation processes (AOPs) for environmental detoxification in the last years, especially for wastewater treatment. The so-called sonophotocatalysis (SPC), the simultaneous use of ultrasound (US) and photocatalysis (PC) by semiconductors to degrade organic pollutants in water (e.g., the effluent of dye works) has been investigated, but combined AOPs methods were not applied to oil-polluted water remediation to our knowledge [17–22].

Among the analytical techniques available for structurally determining crude oil components or metabolites, gas chromatography combined with mass spectrometry (GC-MS) has been the best choice so far and most widely used [23,24]. The fractionation of crude oil and subsequent GC-MS analysis has characterised nearly 300 components comprising aliphatic, aromatic, and biomarker compounds [25–28]. However, most crude oil fractions remain unidentified since many components cannot be resolved and appear as "hump" or "unresolved complex mixture (UCM)" in GC chromatograms [29,30].

Compositions of the saturated hydrocarbons have been better characterised by two-dimensional gas chromatography coupled to mass spectrometry [29] and liquid chromatography-mass spectrometry [31]. However, polar species appear poorly resolved due to their compositional complexity far exceeding the peak capacity of typical analytical techniques. High mass resolving power is necessary for the resolution of many compounds present in crude oil.

The development of Fourier transform ion cyclotron resonance mass spectrometry (FT-ICR MS) had provided the needed ultra-high resolving power (m/$\Delta m_{50\%}$ > 100,000, in which $\Delta m_{50\%}$ is peak width at half peak height), and the use of electrospray ionization (ESI) mass spectrometry had made possible to detect most polar species. Thus, the coupling of these two techniques, ESI and FT-ICR mass spectrometry, produces a powerful analytical tool for analysing these polar species without a preliminary chromatographic separation [32,33].

This work investigates the modifications that the artificial ageing induced on the composition of the polar fraction of an Italian crude oil (Basilicata region—Southern Italy, Val D'Agri countryside) under solar irradiation. Moreover, it explores the possibility of oil-polluted water remediation using AOPs, such as photocatalysis (UV + TiO_2), sonolysis (US, ultrasound irradiation) and the simultaneous use of photocatalysis and sonolysis, i.e., sonophotocatalysis (UV + TiO_2 + US).

2. Results and Discussion

The crude oil sample, collected from the Oil Centre sited in Val D'Agri (Basilicata), underwent simulated solar treatment. Information on the composition of the oil water-soluble fraction obtained through GC-MS, liquid state ^{1}H NMR and FT-ICR-MS was the basis for this investigation. Liquid state ^{1}H NMR spectroscopy accomplished helpful information on the oil composition. This technique recognised amounts of 59%, 19%, and 20% of total hydrocarbons as linear, cyclic (or branched), and aromatic compounds. NMR spectroscopy cannot discriminate branched from cyclic alkanes because both compounds have the same intramolecular environment. Figure S1 compares data obtained by NMR and GC-MS.

The use of real standards introduced by an electrospray ion source allowed the calibration of mass spectra. Recalibration was necessary for the identified homologous series in each sample [34]. A troubling complication in structural studies of crude oil has been its enormous complexity on a molecular scale. The ultrahigh-resolution of FT-ICR spectra can be highly complex: these spectra typically comprise many peaks at each "Nominal mass" and thousands of peaks in a whole spectrum. Each peak could represent a chemically diverse compound. This complexity poses an investigative challenge to the study of spectra for structural interpretation.

The univocal assignment of elementary composition, merely based on the high resolution and accuracy of the instrument, is not possible for all mass values. For values higher than 400–500 Da, it is necessary to validate the result differently. The Kendrick plot (Kendrick mass defect vs Kendrick Nominal mass or KMD vs KNM) offers an outstanding vehicle to visualise and categorise all of the peaks in a mass spectrum. Kendrick mass defect (KMD) breakdown has been effectively applied to ultra-high resolution mass spectra, consenting to categorise peaks into complex spectra based on their homologous similarities across a selected type of masses [35]. Bi-dimensional plots can discern compounds differing by masses associated with a structural unit (e.g., CH_2, COOH, CH_2O, etc.). In this drawing, the signals of structurally related moieties all lie on horizontal or diagonal straight lines. Such a method permits the extraction of peaks that are homogeneously associated. The method can effectively recognise groups of associated compounds in FT-ICR-MS of petroleum samples [36]. The compounds of the same homologous series (having a different number of groups CH$_2$) will fall in a single horizontal line of the diagram (KNM), with peaks separated from 14 Da and no difference of KMD.

Similarly, the signals relating to compounds of the same class but of different types will occupy points on a vertical line of the diagram, separated by a difference of 0.013 in the Kendrick mass defect. The conversion of mass spectra from the IUPAC mass scale (based on the ^{12}C atomic mass as exactly 12 Da) to the Kendrick mass scale is the first step. The Kendrick mass scale poses CH$_2$ = 14.0000 Da rather than 14.01565 Da. The Kendrick mass comes from the IUPAC mass, as shown in Equation (1) [35,36]:

$$\text{Kendrick mass} = \text{IUPAC mass} \times (14.00000/14.01565) \tag{1}$$

Members of a homologous series (specifically, compounds that comprehend the same heteroatom and number of rings plus double bonds, but a different number of CH$_2$ groups) have the same KMDs. They are thus quickly organised and selected from a list of all detected ion masses, as shown in Equation (2):

$$\text{KMD} = \text{KNM} - \text{KEM} \tag{2}$$

where KEM is the Kendrick exact mass.

By rounding the Kendrick mass up to the nearest whole number, the nominal Kendrick mass conveniently arises. Next, homologous series are parted based on even and odd Kendrick Nominal mass and KMD, as described elsewhere [36,37]. Finally, the Kendrick masses are sorted based on Kendrick mass defect and nominal-Z value and exported into an Excel spreadsheet in the second step. Then, a molecular formula calculator programme, limited to molecular formulas consisting of up to ^{12}C 0–80 and ^{16}O 0–10, assigns elemental compositions. Since members of a homologous series diverge only by integer multiples of CH$_2$, the assignment of a single unit of such a series typically suffices to identify all higher-mass members of that series [36].

We also used the van Krevelen diagram for examining ultra-high resolution mass spectra. This kind of layout is used broadly in the geochemistry literature to study the evolution of coals or oil samples [38–40]. The molar hydrogen-to-carbon ratios (H/C) constitute the ordinate, and the molar oxygen-to-carbon ratio (O/C), the abscissa. As a result, each class of compounds plots in a specific location on the diagram. Researchers well recognised that they can identify the type of compounds from the position of their representative points in the van Krevelen plot [41–43].

In general, the chemical formula CcH2c(Z)NnOoSs can identify the crude oil composition. That is because the hydrogen deficiency index <Z> of the molecule is the same for all members of a homologous "type" series (i.e., the fixed number of rings plus double bonds). Every two-units decrease in <Z> value represents the addition of one ring or a double bond. Therefore, number-average molecular weight, Mn, and weight-average molecular weight Mw have a synthetic definition as:

$$Mn = \Sigma iMi/\Sigma Ni \tag{3}$$

and

$$Mw = \Sigma NiMi^2/\Sigma NiMi \tag{4}$$

where Ni is the relative abundance of ions of mass Mi [34].

The <Z> number plays an essential role for the general molecular formula CcH2c(Z)X of the corresponding neutral species, in which X denotes the constituent heteroatom (Nn, Oo, and Ss).

2.1. Ageing Study of Crude Oil by FT-ICR-MS

A solar simulator (Suntest®), equipped with a xenon lamp as the light source used for the ageing treatment, provided information about the crude oil's photochemical behaviour.

GC-MS spectra showed that the fraction present in the highest percentage shifted from the C_8–C_{11} fractions to the C_{13} (Figure S2) in the irradiated sample. We observe an increased amount of the C_{13}–C_{23} and a decreased amount of the C_7–C_{12} fractions. In the natural (not irradiated) oil, the C_8–C_{11} fractions represented 54.8% of all the compounds detected. Figure S2B depicts the distribution of the compounds as a function of their chemical type. The GC-MS analysis of the mixture deriving from solar simulator irradiation showed an increase in the relative amounts of both linear alkanes and aromatic compounds. At the same time, we observed a sharp decrease in the relative amounts of branched chains. After irradiation, we did not find cyclic alkanes and alkenes.

After the irradiation, the compositional analysis of the linear alkanes highlighted several changes (Figure S2C) compared to the not irradiated sample. Undecane was the hydrocarbon found in the highest percentage in the crude oil, while pentadecane was in the irradiated oil. A decrease in the C_7–C_{12} and an increase in C_{13}–C_{25} fractions is evident. All our analytical determinations agree with reducing the number of branched alkanes in crude oil after irradiation. Figure S2D illustrates the modifications of the composition of this fraction. After the irradiation, branched alkanes underwent a sharp reduction and only C_8, C_9, C_{11}, C_{12}, and C_{13} fractions were present. Cyclic alkanes were not present after the solar simulator experiment (Figure S2E).

The percentage area of the aromatic compounds did not vary with solar irradiation (Figure S2B). However, a sharp decrease in benzene-like structures and an increase in naphthalenic ones have been observed (Figure S2F).

Figure 1a,b show the FT ICR-MS spectra of untreated and treated crude oil, respectively. The spectra show the distribution of multiple ions with a single charge comprised between m/z 150 and m/z 1400. Figure 1(c_1–c_3) show the scale-expanded segment of the mass spectrum in Figure 1a, revealing an average period of nominal 14 mass units. The signal intensities increased after light irradiation. The shift of maximum apex was not negligible in the treated crude oil sample, which was also more viscous. Thanks to the high accuracy of mass and the excellent resolution power of FT-ICR-MS, it was possible to carry out the non-ambiguous determination of the elementary composition of multiple isobaric picks.

The chemical formula CcH2c(Z)X generally expresses the composition of a hydrocarbon molecule; where, <c> is the number of carbon atoms, <Z> is the hydrogen deficiency (a measure of aromatic character), and X represents the constituent heteroatom (N, S, O) in the molecule. The heteroatom of interest is oxygen in this study. For simplification, Kendrick and van Krevelen diagrams of natural and irradiated crude oil shown in the figures report only the O_3 class, which contains the most numerous groups of detected ions. Table 1 illustrates an example of homologous series extracted by the mass spectrum of the untreated sample, with a degree of unsaturation Z = −20 and class of oxygen O_3, containing the most numerous groups of detected ions.

The compounds of the same homologous series, having a different number of CH_2 groups, fall in a single horizontal line of the Kendrick plot with peaks separated from 14 Da and no difference in the Kendrick mass defect (Table 1, Figure 2). The compounds of the same class but different typology settle down on a vertical line of the diagram separated from a difference of 0.0134 in the value of KMD. Figure 2 compares Kendrick plots for positive-ion ESI FT-ICR mass spectra of natural (■) and irradiated (X) crude oil samples. Due to the high number of signals, the figure reports only the O_3 class, containing the most numerous group of detected ions. Kendrick plot of crude oil sample for O_3 class shows many compounds with a high degree of unsaturation (high value of KMD). In the low values of KMD, the highest percentage of compounds has a small alkylation series (limited number of -CH_2- moieties).

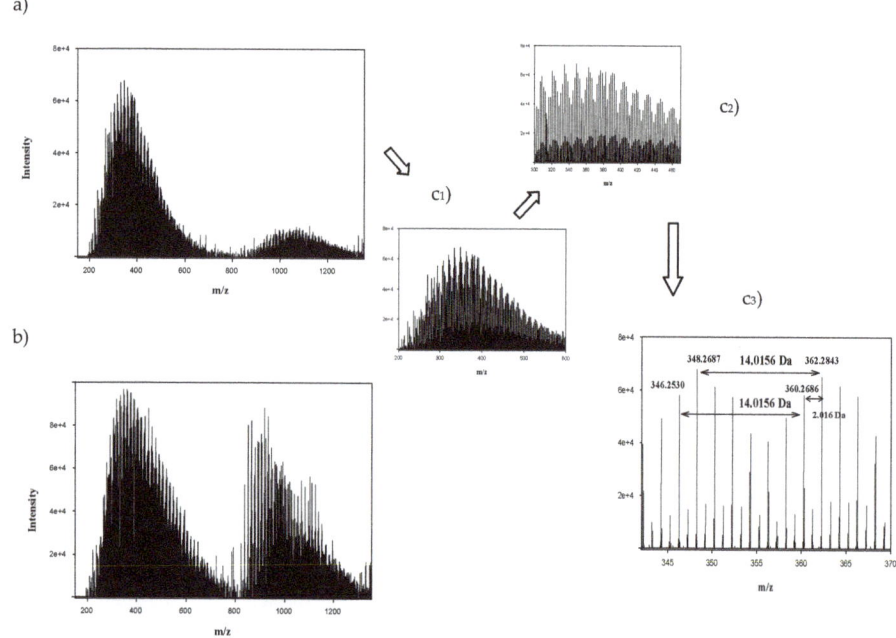

Figure 1. (**a**) FT ICR-MS spectrum of the untreated oil sample; (**b**) FT ICR-MS spectrum of the same sample irradiated by xenon-lamp. The spectrum shows the distribution of multiple ions with a single charge comprised between m/z 150 and 1400; c_1, c_2, c_3 insets = mass scale-expanded segments of the full range crude oil mass spectrum in Figure 1a, revealing periodicities of 14.016 Da from compound series differing in the number of CH_2 groups and 2.016 Da from compound series differing in the number of rings plus double bonds.

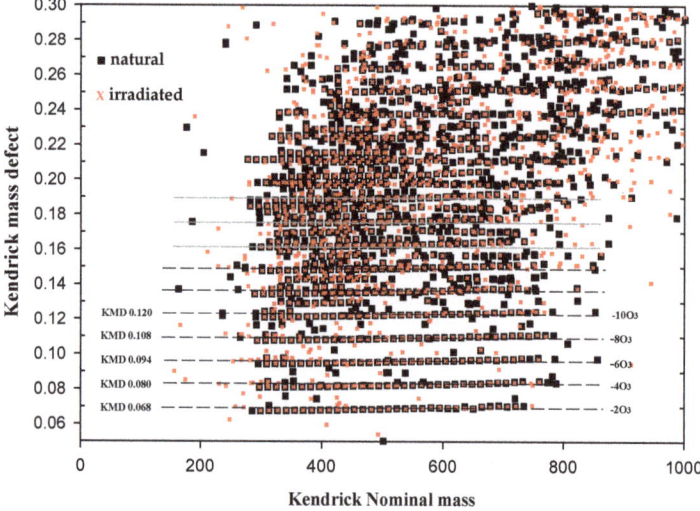

Figure 2. Kendrick mass plot of the O_3 species found in natural (■) and irradiated oil (X). This plot illustrates the increase in the number of rings plus double bonds as the KMD increases (y-axis) and the alkylation series along the x-axis.

Table 1. Homologous series of O_3 class with $<Z> = -20$.

m/z	Intensity	Composition	KNM	KMD
337.1818	3941.7	$C_{22}H_{25}O_3$	4721	0.45494
351.1972	5150.2	$C_{23}H_{27}O_3$	4917	0.23920
365.2128	3112.4	$C_{24}H_{29}O_3$	5113	0.02038
379.2270	4609.2	$C_{25}H_{31}O_3$	5310	0.82256
393.2424	4252.4	$C_{26}H_{33}O_3$	5506	0.60640
407.2582	3311.8	$C_{27}H_{35}O_3$	5702	0.38464
421.2737	3427.8	$C_{28}H_{37}O_3$	5898	0.16848
435.2893	4966.3	$C_{29}H_{39}O_3$	6095	0.94924
449.3052	3587.2	$C_{30}H_{41}O_3$	6291	0.72748
463.3202	4252.1	$C_{31}H_{43}O_3$	6487	0.51692
477.3362	4772.3	$C_{32}H_{45}O_3$	6683	0.29390
505.3668	7455.6	$C_{34}H_{49}O_3$	7076	0.86536
533.3981	5506.9	$C_{36}H_{53}O_3$	7468	0.42618
547.4131	6265.8	$C_{37}H_{55}O_3$	7664	0.21674
561.4292	6183.2	$C_{38}H_{57}O_3$	7861	0.99120
575.4446	5847.4	$C_{39}H_{59}O_3$	8057	0.77504
589.4606	5685.7	$C_{40}H_{61}O_3$	8253	0.55118
603.4760	6234.8	$C_{41}H_{63}O_3$	8449	0.33586
617.4921	5529.1	$C_{42}H_{65}O_3$	8645	0.11018
631.5075	3522.3	$C_{43}H_{67}O_3$	8842	0.89486
645.5230	4399.5	$C_{44}H_{69}O_3$	9038	0.67870
659.5385	5131.9	$C_{45}H_{71}O_3$	9234	0.46086
673.5545	4308.2	$C_{46}H_{73}O_3$	9430	0.23700
687.5703	4191.7	$C_{47}H_{75}O_3$	9626	0.01566
701.5854	2133.1	$C_{48}H_{77}O_3$	9823	0.80454
715.6014	3178.9	$C_{49}H_{79}O_3$	10,019	0.58068
729.6165	3507.6	$C_{50}H_{81}O_3$	10,215	0.36872
743.6323	3063.6	$C_{51}H_{83}O_3$	10,411	0.14836

This plot can visually sort up to thousands of compounds horizontally according to the number of CH_2 groups and vertically according to class (heteroatom composition) and type (rings plus double bonds). Since these two classes have the same number of oxygen atoms, they have identical O/C ratios but distinguish themselves by different H/C ratios.

The attained results elucidate the transformation of oil components following irradiation. After irradiation with the xenon lamp (Suntest®), a slight shift of the peak to the higher masses appears in the recorded mass spectra, according to Griffiths et al. findings [31]. Therefore, it seems that a phenomenon of molecular polymerisation prevails on the destruction of the tri-, tetra- and penta-aromatic groups. Furthermore, since the increase in unsaturation correlates with the higher toxicity [44], our results could indicate higher toxicity for the oil after irradiation.

The plot of Figure 2 highlights the increase in the number of double bonds' rings as the KMD increases (y-axis) and the alkylation series along the x-axis. The solar irradiation causes a diminution of rings or double bonds (picks rarefaction in samples irradiated), a consequent Kendrick Nominal mass raising of 2 Da, and the Kendrick mass defect diminution. The irradiated crude oil sample shows an expansion of alkylation in compounds with a high degree of unsaturation and a reduced unsaturation number for molecules with a low alkylation degree.

Figure 3 shows the van Krevelen plot for the class of O_3 compounds found in the natural crude oil. The compounds in homologous series, corresponding to varying degrees of alkylation, appear along lines that intersect the value of 2 on the H/C axis. Similarly, a vertical line connects homologous series differing in degree of unsaturation. In agreement with the results in the Kendrick plot, most compounds have a low number of oxygen atoms and a high degree of unsaturation.

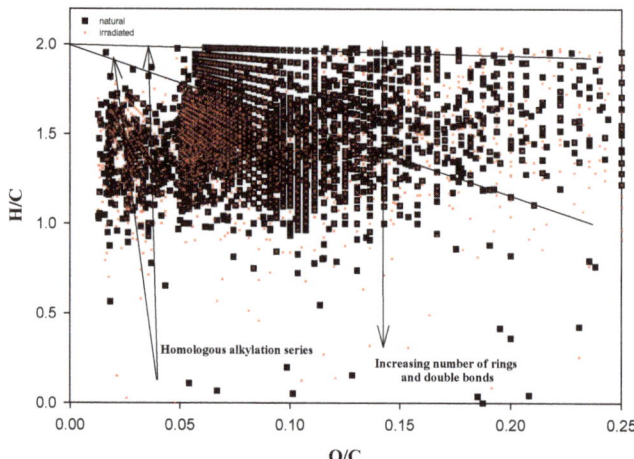

Figure 3. van Krevelen plot of the O_3 species found in natural (■) and irradiated oil (**X**). The compounds in homologous series, corresponding to varying degrees of alkylation, appear along lines that intersect the value of 2 on the H/C axis. Similarly, a vertical line connects homologous series differing in degree of unsaturation.

As the H/C ratio increases, the number of rings plus double bonds decreases. Thus, a slight shift to a lower H/C ratio (i.e., a higher number of rings plus double bonds) occurred. Figure 3 shows a minor shift of the data to the right due to increased oxidation and slight dehydration (the picks shift to the lower left) of hydrocarbons. Kendrick mass defect analysis has dramatically facilitated the interpretation of mass spectra, but it is still challenging to derive details for molecules that contribute to complex ultrahigh-resolution mass spectra.

Figure 4 shows the distribution of compounds associated with their number of oxygen atoms in natural and irradiated samples. In both samples, the number of total oxygenated compounds increases. The augmentation of oxygenated compounds should mainly refer to the O_3 and O_4 types present in the investigated model. The irradiation of crude oil in the solar simulator produces oil oxidation with a particular effect of double bonds oxygenation.

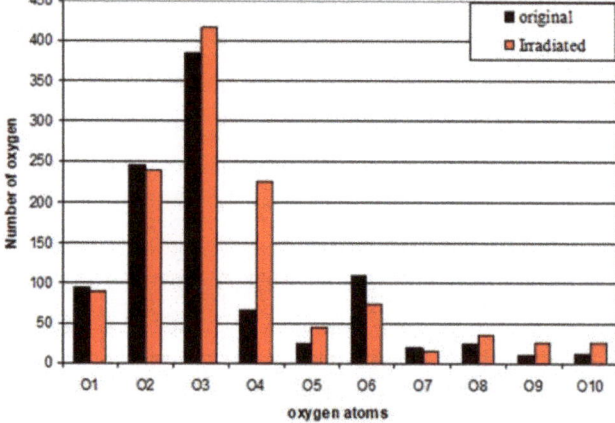

Figure 4. FT-ICR compositional analysis of natural and irradiated crude oil samples as a function of the number of oxygen atoms.

Figure 5 shows oxygen class Z-distributions for natural and irradiated samples, confirming a diminution of hydrogen deficiency index (~15–30% less) and augmentation of oxygen number after the light irradiation. Therefore, the light irradiation induces a manifest photo-oxidation of the crude oil composition. These results highlight toxicity as most of the new oxidised compounds are water-soluble, available in higher concentrations to the living organisms and probably more reactive and biologically active than their parent compounds [43,44].

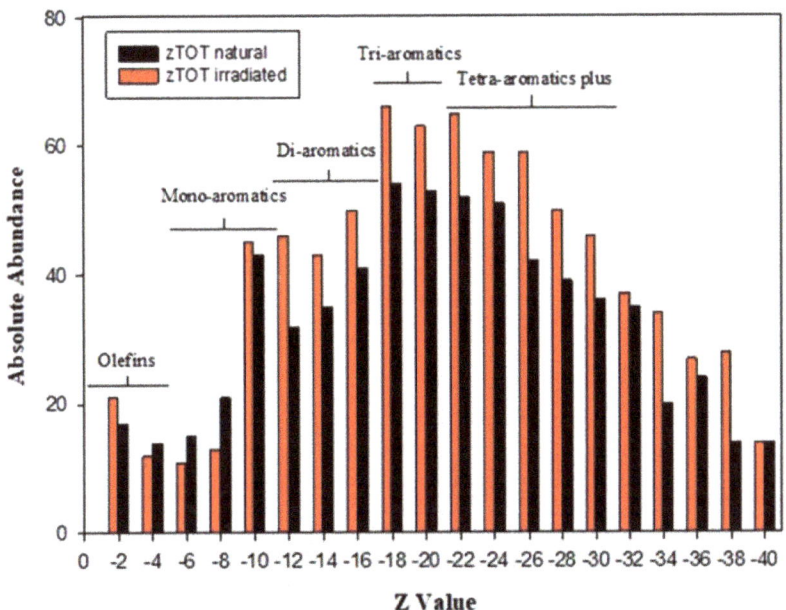

Figure 5. Oxygen class <Z>-distributions for natural and irradiated samples.

2.2. Remediation of Oil-Polluted Water

Since crude oil lies over the surface of water and soil, it suffers solar irradiation. Solar degradation is one of the natural ways for petroleum decontamination, and, as a consequence, techniques based on light irradiation could be advantageous in the petroleum degradation processes. Enhanced light irradiation-based technologies are available, adopting different approaches for the scope [10–12,45].

The accidental dispersion of crude oil in water bodies forms a characteristic thin layer of not water-miscible compounds and a deeper layer of solubilised substances, which cannot easily separate from the aqueous solvent. In this direction, our approach was to prepare a water/oil suspension and investigate the efficiency of different cleaning methods. The water-soluble fraction of crude oil was undergone degradation by photocatalysis, sonolysis, and sonophotocatalysis, i.e., the simultaneous use of UV, titanium dioxide, and ultrasound emitter (UV + TiO_2 + US). GC-MS, liquid-state NMR, fluorescence, and high-resolution mass spectrometry (FT-ICR) analyses elucidated the chemical nature of water-soluble organic compounds after degradation processes and liquid-liquid extractions (LLEs). The results obtained in this study are concisely readable in Table 2.

Table 2. Synthetic results obtained from the different photodegradation processes of crude oil and oil water-soluble fraction (WSF) under investigation.

Degradation Method/System	GC-MS	^{1}H-NMR	FT-ICR MS	Fluorescence
Photolysis (UV)/ Ageing of crude oil	- increase of C_{13}–C_{23} classes and a decrease of C_7–C_{12} types of compounds		- signal intensities increase-augmentation of compounds with a low molecular weight - a light increase of the number of oxygen atoms in the oxygenated species, in particular, O_3 and O_4 types - oxidation of crude oil with a particular effect on double bonds' oxygenation	
Photocatalysis (UV + TiO_2)/ Oil Water-Soluble Fraction (WSF)	- increase of C_5 compounds from 67% to 89% - decrease of C_6, C_7, C_8 and C_9 compounds - increase of branched alkanes from 50% to 65% - increase of cyclic alkanes from 4% to 5% - decrease of aromatic compounds from 23% to 13% - decrease of linear alkanes from 22% to 14%	- slight increase of linear and cyclic alkanes and a sharp decrease in aromatics - no other significant differences emerged in the composition of WSF before and after the processes	- a slight decrease in the total number of oxygenated compounds - the O_1 and O_2 classes prevailed over the other types	- aromatic compounds' decrease (about 46% after 1 h of treatment)
Sonolysis (US)/ Oil Water-Soluble Fraction (WSF)	- increase of C_5 compounds from 67% to 91% - disappeared C_9 class - increase of branched alkanes from 50% to 54% - increase of cyclic alkanes from 4% to 5% - increase of aromatic compounds from 23% to 24% - decrease of linear alkanes from 22% to 17%	- no significant differences emerged in the composition of treated and not treated WSF	- a low increase of oxygenated compounds: O_1, O_2 and O_7 classes and decrease of the other oxygenated types - a sharp increase in the number of compounds with a low molecular weight	- no significant differences emerged in the composition of aromatic compounds before and after the processes
Sonophotocatalysis (UV + TiO_2 + US)/ Oil Water-Soluble Fraction (WSF)	- increase of C_5 compounds from 67% to 91% - decrease C_6, C_7 and C_8 compounds; - disappeared C_9 class - increase of branched alkanes from 50% to 64% - increase of cyclic alkanes from 4% to 9% - significant decrease of aromatic compounds from 23% to 7% - decrease of linear alkanes from 22% to 19%	- significant decrease of aromatic compounds from 23% to 7%	- the total number of oxygenated compounds decreased from 1203 to 993 - increase of compounds with low molecular weight and compounds with a low unsaturation degree	- aromatic compounds' decrease (about 48% after 1 h of treatment)

2.2.1. Photocatalytic Degradation

In the photocatalytic process, the water/oil suspension was treated for 1 h with UV irradiation in the presence of titanium dioxide. GC-MS analysis of WSF (Figure S3) evidenced increased C_5 compounds from 67% in not-treated WSF to 89% in the irradiated sample. Moreover, the amount of C_6, C_7, C_8, and C_9 compounds decreased. The analysis of chemical classes occurring in the irradiated WSF showed increased branched and cyclic alkanes, from 50% to 65% (branched) and from 4% to 7% (cyclic), respectively. On the other hand, the number of linear alkanes underwent a slight decrease (from 22% to 14%), and the aromatic compounds had a sharp decline (from 23% to 13%).

^{1}H-NMR spectra (Figures S4 and S5) confirmed a slight increase of linear and cyclic alkanes, and a sharp decrease in aromatics, as evidenced in the chromatographic analysis.

FT-ICR MS analysis showed a minor decrease in the total number of oxygenated compounds. The O_1 and O_2 classes prevailed over the other types (Figure 6).

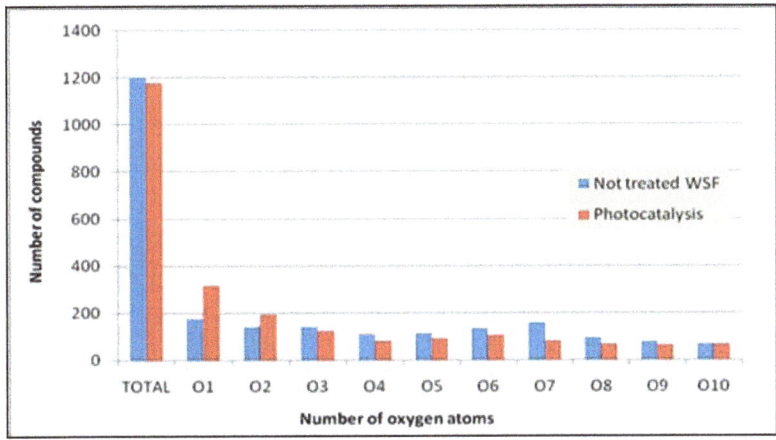

Figure 6. FT-ICR MS analysis of natural and 1-h photocatalysed WSF of crude oil as a function of the number of oxygen atoms.

Comparison of Kendrick plots constructed for the untreated sample (Figure 7a) and the photodegraded model (Figure 7b) shows an increase in the number of compounds with low molecular weight and low degree of unsaturation. The formation of several homologous series, with KDM values of 0.124, 0.137, 0.150, and so on, is underlined in the Kendrick diagram plotted for the treated sample. The unsaturation degree of these homologous series falls in the range Z = −16 to Z = −20.

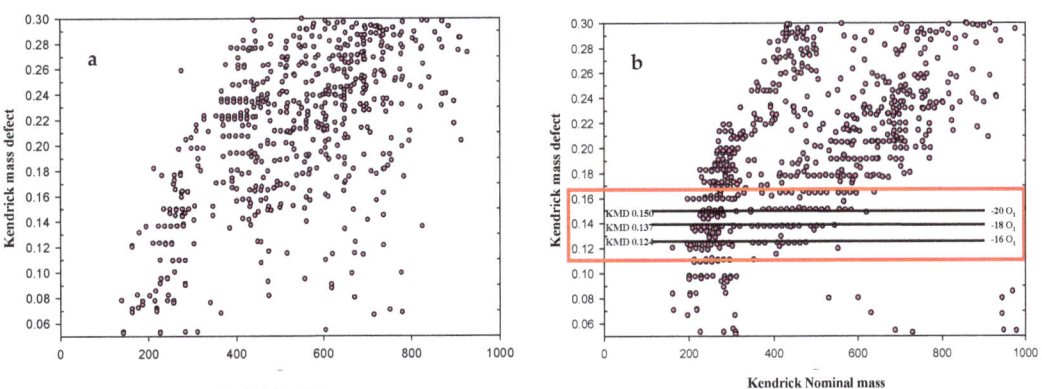

Figure 7. Kendrick mass plots of the O_1-O_{10} species found in the untreated crude oil WSF (**a**) and the 1-h photocatalysed sample (**b**).

The van Krevelen diagram (Figure 8) shows an increase in O_1 class, reduced O/C ratio, and decreased unsaturated compounds in the treated sample compared to the untreated one. After photocatalysis, the number of compounds with a low number of oxygen atoms increased. As shown in Figure 6, the O_1 and O_2 classes prevailed over the other types, resulting in a decrease in the O/C ratio.

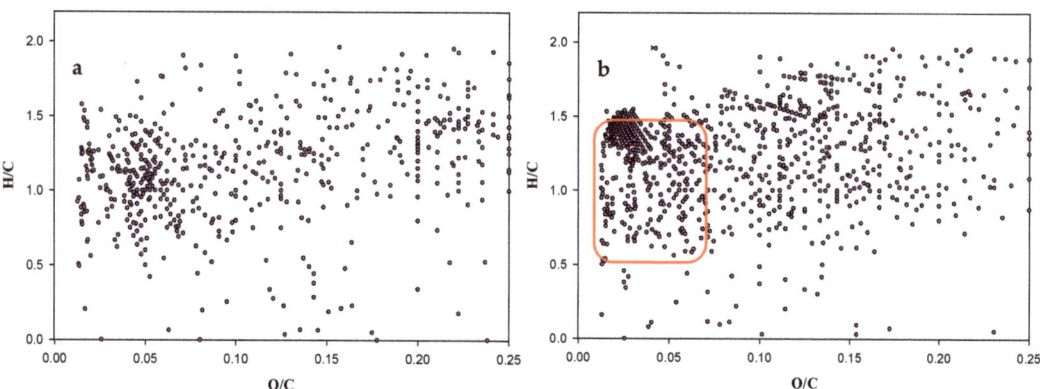

Figure 8. The van Krevelen plots of the O_1–O_{10} species found in the untreated crude oil WSF (**a**) and the 1-h photocatalysed sample (**b**).

From fluorescence spectra (Figure S6), it was possible to argue aromatic compounds' decrease after 1-h photocatalytic treatment. Essentially, the absolute intensity of the peak at 347 nm decreases from 92.07 mAU for the natural sample to 49.50 mAU for the treated sample, with a reduction of 46%.

2.2.2. Ultrasonic Irradiation

In the sonolytic process, the water/oil suspension received 1-h ultrasound irradiation. GC-MS analysis (Figure S7) indicated that C_5 compounds increased from 67% to 91% at the end of sonolysis, evidencing a behaviour analogous to photocatalysis. Furthermore, the C_6, C_7 and C_8 compounds decreased, similarly to the photocatalytic process; otherwise, the C_9 class disappeared. The analysis of functional groups evidenced that branched alkanes increased from 50% to 54% in the sonolysed WSF (from 50% to 65% in photocatalysis). Cyclic alkanes underwent a minor increase from 4% to 5% (like photocatalysis), but aromatic compounds slightly increased from 23% to 24% (decreased dramatically to 13% in photocatalysis). The number of linear alkanes decreased from 22% to 17% (22% to 14% in photocatalysis).

Liquid-state ^{1}H-NMR spectra (Figures S8 and S9) evidenced a relatively equal amount of the three classes of compounds in the not-treated and sonolysed samples. In conclusion, no significant differences emerged in the composition of WSF before and after the processes of sonication and photocatalysis, with a unique exception for aromatic compounds, as mentioned above in the case of photocatalysis.

From FTICR MS analysis, the total number of oxygenated compounds registered a low increase in the sonicated sample. O_1, O_2 and O_7 classes increased, but the other oxygenated types decreased (Figure 9).

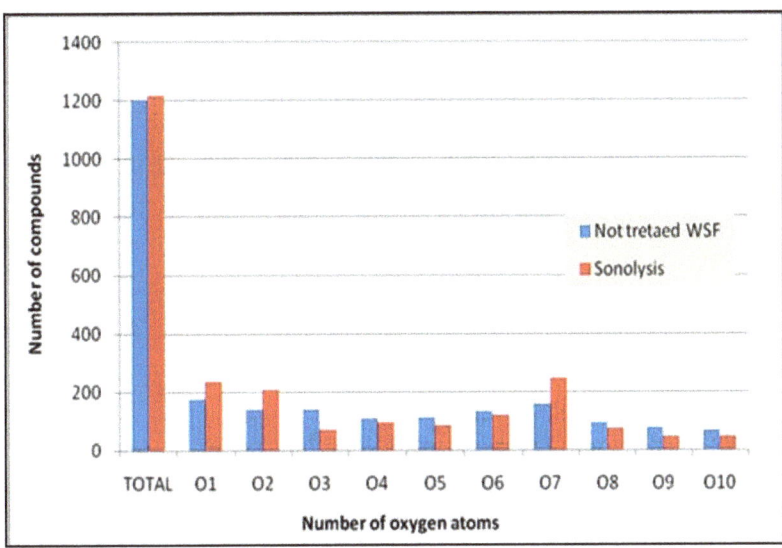

Figure 9. FT-ICR analysis of natural and 1-h sonicated WSF of crude oil samples as a function of the number of oxygen atoms.

Analysis of the Kendrick plot (Figure 10) after the degradation treatment shows a sharp increase in the number of compounds with low molecular weight. Seventy percent of compounds stay in the range m/z 159–597, and many homologous series with a high degree of unsaturation are visible.

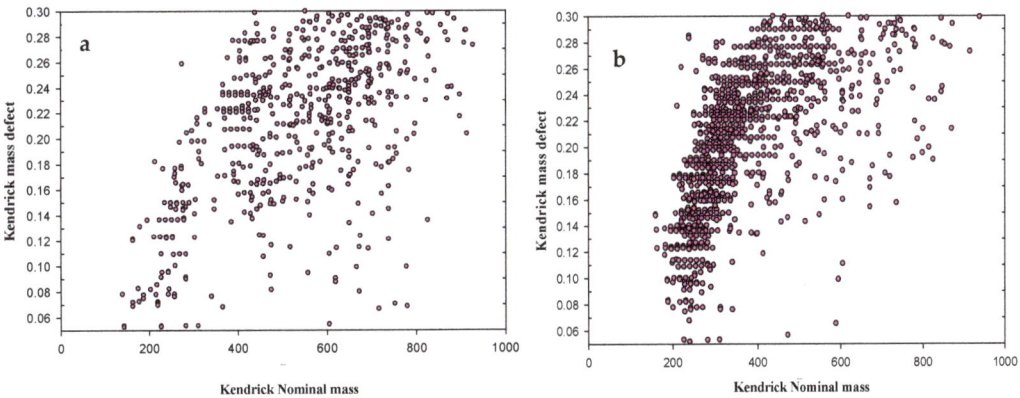

Figure 10. Kendrick mass plot of the O_1–O_{10} species found in the not-treated WSF of crude oil (**a**) and after 1-h of US treatment (**b**).

The van Krevelen diagram (Figure 11) substantiates any differences between the natural and treated WSF samples.

The fluorescence study (Figure S10) confirms that the decrease of aromatic compounds is not so evident with US treatment. After 1-h of the sonolytic process, the absolute intensity of the maximum peak displays an insignificant drop from 99.96 mAU for the natural sample to 93.37 mAU for the treated one.

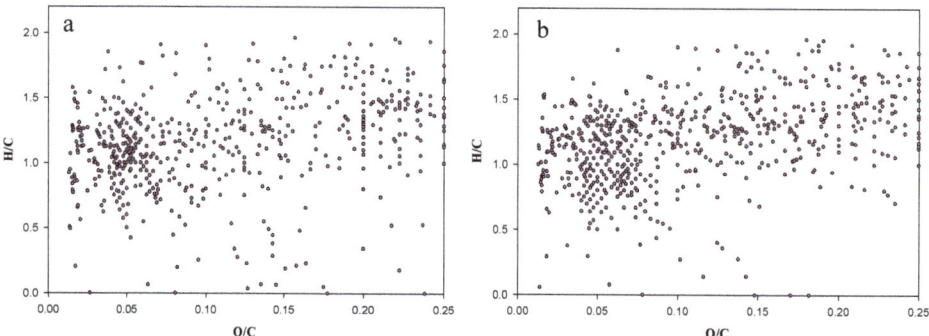

Figure 11. The van Krevelen plots of the O_1-O_{10} species found in the not-treated WSF of crude oil (**a**) and after 1-h US treatment (**b**).

2.2.3. Sonophotocatalytic Degradation

The contemporary use of UV irradiation, titanium dioxide and ultrasound irradiation to treat the oil aqueous suspension shows results mainly similar to those obtained with sonolysis or photocatalysis.

GC-MS analysis (Figure S11) demonstrated that after 1-h of treatment, C_5 compounds increased from 67% in not-treated WSF to 91% in the treated sample, while C_6, C_7 and C_8 compounds decreased; C_9 compounds were not detected (like the simple US). The analysis of functional groups in the sonophotocatalytic degradation evidenced an increase from 50% to 64% of branched alkanes (similar to photocatalysis) and from 4% to 9% of cyclic alkanes (higher than the other technologies). The number of linear alkanes underwent a slight decrease (from 22% to 19%, similar to the other technologies). In comparison, aromatic compounds showed the sharpest decline (from 23% to 7%), also proved by integrating NMR spectra (Figures S12 and S13). In the natural WSF, the aromatics alkanes occupied 19% of the whole spectral area, whilst in the treated sample, this amount decreases up to 3.3%. On the other hand, the amount of linear and cyclic alkanes increases by about 7–8%.

Figure 12 shows the trend of oxygenated compounds after sonophotocatalytic treatment. In this case, the total number of oxygenated compounds decreased from 1203 (not treated WSF) to 993 (treated WSF).

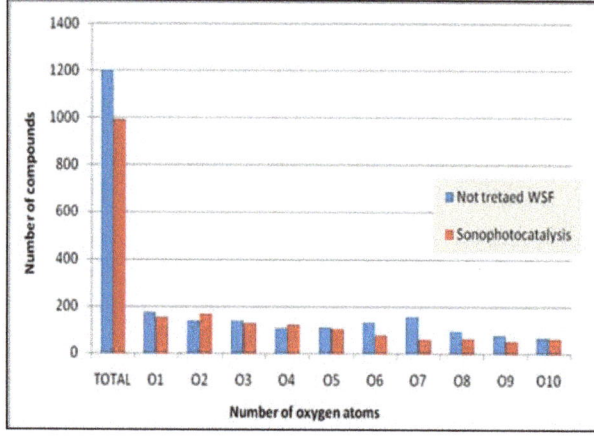

Figure 12. FT-ICR MS analysis of natural and 1-h sonophotocatalysed WSF sample of crude oil as a function of the number of oxygen atoms.

Comparison of Kendrick plots (Figure 13) obtained for the not-treated and treated samples showed an increased number of compounds with low molecular weight, especially in the range m/z 169–369, and compounds with a low unsaturation degree.

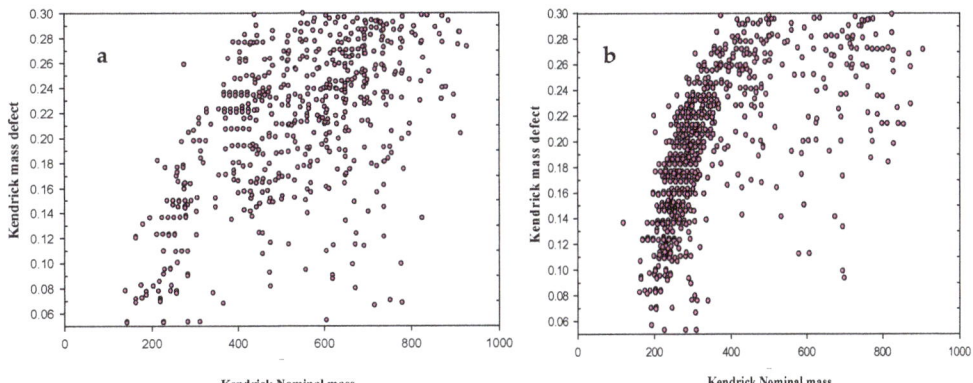

Figure 13. Kendrick mass plots of the O_1-O_{10} species found in the not-treated WSF of crude oil (**a**) and after 1-h of sonophotocatalytic treatment (**b**).

The van Krevelen diagram (Figure 14) let us see an intensification of signals relative to oxygenated compounds with an O/C ratio in the range 0.10–0.25.

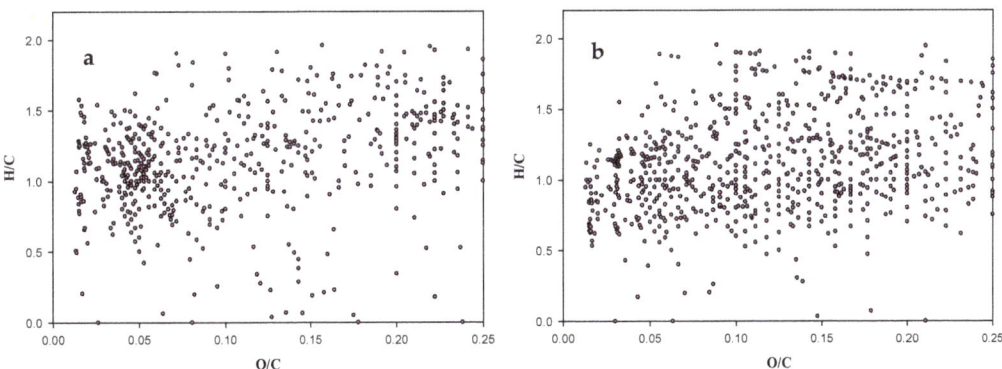

Figure 14. The van Krevelen plots of the O_1-O_{10} species found in the not-treated WSF of crude oil (**a**) and after 1-h of sonophotocatalytic treatment (**b**).

Fluorescence spectra (Figure S14) substantiated the decreasing of aromatic compounds after 1-h sonophotocatalytic treatment. The absolute intensity of the peak at 347 nm decreased from 89.92 mAU for the natural sample to 47.95 mAU for the treated sample, reducing by 48%.

3. Materials and Methods

3.1. Crude Oil and Chemicals

The director of the Eni-Cova Oil Plant in Val d'Agri (Basilicata Region, Southern Italy) kindly provided the oil sample taken from the first step of oil purification after extraction, including dehydration and degasification. Table 3 accounts for the elemental composition reported in the label accompanying the sample delivered for this research.

Table 3. Elemental composition (%) of the oil sample taken from the first step of oil purification after extraction, including dehydration and degasification [a].

C	H	N	O [b]	S
85.4 ± 2.8	11.8 ± 1.4	0.3 ± 0.1	0.6 ± 0.4	1.9 ± 0.9

[a] Metals (Ni and V) < 1000 ppm. [b] Obtained as the complement to 100.

All chemicals used were of analytical grade. TiO_2 Degussa P-25, obtained as a gift from Evonik (Hanau, Germany), was the catalyst adopted. Table 4 reports a summary scheme of the investigation executed.

Table 4. Experiments performed and analytical methods used in this study.

Experiments Performed	Treated System	Analytical Methods Applied			
		GC-MS	^{1}H-NMR	FT-ICR MS	Fluorescence
Artificial ageing process by photolysis (UV)	Crude oil	X		X	
Remediation process by photocatalysis (UV + TiO_2)	Oil WSF	X	X	X	X
Remediation process by sonolysis (US)	Oil WSF	X	X	X	X
Remediation process by sonophotocatalysis (UV + TiO_2 + US)	Oil WSF	X	X	X	X

3.2. Photodegradation Apparatus

The Suntest CPS+ (Heraeus Industrietechnik GmbH, Hanau, Germany), equipped with a xenon lamp of 1.1 kW, protected employing a quartz plate (total passing wavelength: 300 nm < λ < 800 nm), was the solar simulator adopted for photochemical reactions. The temperature of the irradiation chamber was 25 °C, maintained through both a thermostatic bath and a conditioned airflow. During the experiments, the crude oil samples were kept up in the horizontal position, creating a homogeneous film of 0.5 cm thickness.

3.3. Photodegradation Process and Sample Preparation for ESI FT-ICR MS

The protocol used for oil ageing experiments was: (i) the irradiation of the natural crude oil (10 mL) for a week in the borosilicate planar reactor; (ii) crude oil samples preparation by dissolving ~30 mg of material in 30 mL of toluene; (iii) withdrawal of 1 mL solution and its dilution with 0.5 mL methanol; addition of either 10 µL acetic acid (for positive ion ESI) or 10 µL ammonium hydroxide (for negative ion ESI) to facilitate protonation or deprotonation in the electrospray ionisation process, respectively.

3.4. Ultrasonic Irradiation of WSF Samples

A crude oil/water suspension was arranged in a borosilicate decanter (5 L) equipped with a Teflon tap at the bottom. The decanter was filled with 3.5 L of ultrapure water; the crude oil was added at the ratio of 1/20 (oil/water), and the suspension was magnetically stirred and then kept in the dark for 30 days at constant temperature (25 °C) to reach equilibrium and the separation of oil phase on the surface of the aqueous phase. Aqueous samples were drawn off through the Teflon tap without disturbing the oil/water separation surface. The collected aqueous sample (500 mL) underwent cotton filtration, to avoid the formation of an emulsion in the solution. The ultrasonic degradation tool was the immersible ultrasonic emitter Sinaptec Nexus P198-R (Sinaptec, Lezennes, France), an ultrasonic module furnished with a titanium sonotrode (S23-10-1/2, Sinaptec), an electrical signal of frequency close to 20 kHz, and a voltage of about 1 kV. In this configuration, the electric power provided by the generator (Nexus P198-R, same manufacturer) is adjustable between 7 W and 100 W, as indicated on a digital display panel. However, this electrical measurement does not determine with high precision the acoustic power dissipated in the liquid. The experimental temperature was fixed to 25 °C.

3.5. Photocatalytic and Sonophotocatalytic Degradation of WSF Samples

The photocatalytic method to degrade the water-soluble fraction of crude oil utilises a 125 W high-pressure mercury lamp (Philips-HPK, Philips, Turnhout, Belgium) that provides its maximum energy at 365 nm, with a range of emission from 195 to 580 nm. The catalyst was TiO_2 (80% anatase–20% rutile). The simultaneous use of the mercury lamp, titanium dioxide and the ultrasound emitter (UV + TiO_2 + US) permitted the sonophotocatalytic degradation. The experimental temperature was 25 °C for photocatalysis and sonophotocatalysis trials.

3.6. Liquid–Liquid Extraction (LLE)

The experimental protocol was (i) to collect samples of the oil WSF after 15, 30, 45, and 60 min of treatment; (ii) to extract in triplicate 30 mL of each sample in a separatory funnel (50 mL) with 3 mL dichloromethane for GC-MS analysis and (iii) another 30 mL with the same solvent for ^{1}H-NMR spectroscopy; (iv) to perform fluorescence analysis using 5 mL of the aqueous solution without extraction.

The internal standard used for assessing the reproducibility of WSF extraction was 1 mL of 1,3-dibromopropane (26.7 mg L^{-1}) added to the volume of dichloromethane needed for each liquid–liquid extraction. In addition, it was necessary to add 1.0 mL of 1-bromododecane (29.0 mg L^{-1}) to evaluate the GC-MS analysis reproducibility at the end of each extraction. Thus, the injection volume was 1 µL extract for each GC-MS run.

3.7. Analysis of Fluorescence

A research-grade spectrofluorometer FP-6500 Jasco (Jasco Corporation, Cremella, Italy) was available for fluorescence analysis. This analysis was necessary to appreciate the aromatic compounds remaining in the water-soluble fraction of crude oil. The spectrofluorometer FP-6500 Jasco, adopting as emitting source a DC-powered 150 W xenon lamp (in a sealed housing), employs a photometric rationing system, which utilises a second photomultiplier tube to monitor and compensate for any variations in the intensity of the xenon source, thus ensuring maximum analytical stability. Furthermore, a concave holographic grating monochromator with optimised blaze angles provides maximum sensitivity over the entire wavelength range; (220–750 nm; 1 nm resolution).

The fluorescence optical path adopted was 1 cm in quartz cells (volume ca 5 cm^3) at 237 and 320 nm excitation and 347 and 360 nm emission.

3.8. ^{1}H-NMR Analysis

A Varian Oxford AS400 spectrometer (Palo Alto, CA, USA), operating at 400 MHz, was enough for the ^{1}H-NMR spectra recording. The set temperature for the used 5 mm non-gradient broadband inverse (BBI) probe was 25 °C.

All the ^{1}H-NMR spectra have tetramethylsilane (TMS) as reference under the acquisition parameters shown in Table 5.

In degradation experiments, liquid–liquid extraction with chloroform permitted to isolate the water-soluble fraction of crude oil. After the complete evaporation of the solvent in a rotary evaporator, the addition of 500 µL deuterated chloroform ($CDCl_3$) permitted to recuperate the residual organic mixture.

Table 5. ^{1}H-NMR acquisition parameters.

Instrument	Solvent	Acquisition Time	Spectral Width	Line Broadening	Number of Scans
Varian Oxford AS400	CDCl$_3$	2.049 s	6410.3 Hz	0.20 Hz	512

3.9. Mass Spectrometry of Polar Components

The instrument available to determine polar components was the micro ESI/FT-ICR/MS 7 T Thermo Electron (Waltham, MA, USA).

The method used for the routine analyses permitted a mass accuracy better than 2 ppm by external calibration, using the mixture of caffeine, MRFA, and Ultramark. The technique separated more than 6000 ion signals belonging to chemically different elemental compositions with a 200,000 resolving power (m/Δm$_{50\%}$ at m/z 400) in positive electrospray mode. The robustness of this equipment, combined with unprecedented ease of use, ultra-high mass accuracy, high sensitivity, and excellent resolving power, make it an ideal instrument for analysis. The infusion of the samples at a flow rate of 5 µL/min permitted the best result in terms of spectrum resolution. ESI conditions were: needle voltage, +4.5 kV; heated capillary current 4 A; tube lens voltage 135.12; temperature 300 °C; N$_2$ speed 2.33 u.a.; aux gas flow rate 0.73; scansions per second 1000.

4. Conclusions

In this study, we tried to characterise the ageing process of crude oil simulating solar irradiation on a thin layer of an oil sample. As a result, FT-ICR MS evidenced an augmentation of compounds with low molecular weight and a slight increase of the number of oxygen atoms in the oxygenated species, as depicted in Kendrick and van Krevelen diagrams. Furthermore, the simulated ageing produced the oxidation of crude oil with a particular effect on double bonds' oxygenation, as confirmed by the disappearance of alkenes in gas chromatographic analysis. The observed results seem to be recognisable because the energy irradiated with the xenon lamp could be enough for catalysing the reaction of olefins with the atmospheric oxygen following a bridge mechanism.

We experimented with different solutions for the cleaning treatment of oil-polluted water (photocatalysis, sonolysis, and sonophotocatalysis). GC-MS analyses of the water-soluble fraction of crude oil for both natural and treated samples discovered that only a few compounds are detectable in the aqueous solution, principally C$_5$-organic chains (~50%). Low amounts of C$_6$, C$_7$, C$_8$ and C$_9$ chains were also present. Both GC-MS and liquid state ^{1}H-NMR signals showed that the branched alkanes were the principal chemical class in the soluble fraction of oil, followed by a small amount of linear and aromatic alkanes.

With all the degradation methods utilised, an increase of the C$_5$-class and a decrease of C$_6$–C$_9$ types of compounds was evident. Furthermore, the FT-ICR comparative analyses of oxygenated species elucidated that the total number of O-compounds in the treated WSF samples is different for all of the degradation methods experimented. The number of the oxygenated compounds slightly decreased with photocatalysis compared to the non-treated sample. An opposite trend appeared with the sonolysis treatment. The sonophotocatalytic method showed a sharp reduction in the number of oxygenated compounds, probably due to the volatilisation of small molecules formed during the oxidation process. It is conceivable that ultrasound can promote this volatilisation. The degradation of the water-soluble fraction of crude oil performed with photocatalysis and sonophotocatalysis led to an apparent decrease of aromatic compounds of 46% and 48%, respectively, for the two techniques, as also confirmed by the fluorescence analysis. With the use of sonolysis, there was no effect on the number of aromatic compounds. Nevertheless, all the degradation methods applied were capable of increasing the number of cyclic alkanes. Therefore, we could speculate that ultrasound in sonophotocatalytic technology can affect the rate of the photocatalytic degradation of the organic pollutants due to a synergistic effect typically observed with an increase of the degradation process efficiency.

In conclusion, our results confirm the photo-oxidation effect caused by light irradiation either on crude oil (simulated ageing) or on the soluble oil fraction. Naturally, the behaviour of each oil type could be different, and then it is not possible to generalise our findings to all cases of oil spilling and environmental remediation. Therefore, it is necessary to check case by case before reaching specific solutions for more efficient remediation processes to avoid making the situation worse.

Supplementary Materials: The following are available online at https://www.mdpi.com/article/10.3390/catal11080954/s1 Figure S1: Percentage of compounds in crude oil as recognized by GC-MS (blue column) and ^{1}H-NMR (red column). Figure S2: GC-MS compositional analysis of crude oil (blue column) and solar simulator irradiated crude oil (red column), as a function of the number of carbon atoms (A); composition of crude oil as a function of the type of compounds, LH: linear aliphatic hydrocarbons; BH: branched aliphatic hydrocarbons; CH: cyclic aliphatic hydrocarbons; AH: aromatic hydrocarbons; AL: alkenes (B); composition of the linear aliphatic hydrocarbons as a function of the number of carbon atoms (C); composition of the branched aliphatic hydrocarbons fraction as a function of the number of carbon atoms (D); composition of the cyclic hydrocarbons as a function of the number of carbon atoms (E); composition of the aromatic hydrocarbons fraction as a function of the number of carbon atoms (F). Figure S3: GC-MS compositional analysis of WSF crude oil before (red column) and after (blue column) photocatalysis: distribution of hydrocarbons as a function of the number of carbon atoms (A) and distribution of the compounds as a function of chemical species (B). Figure S4: ^{1}H-NMR spectra of WSF crude oil before (A) and after (B) photocatalysis. Figure S5: ^{1}H-NMR compositional analysis of WSF crude oil before (red column) and after (blue column) photocatalysis: distribution of the compounds as a function of chemical species. Figure S6: Fluorescence spectra of WSF crude oil before (blu line) and after (red line) photocatalysis. Figure S7: GC-MS compositional analysis of WSF crude oil before (red column) and after (blue column) sonolysis: distribution of hydrocarbons as a function of the number of carbon atoms (A) and distribution of the compounds as a function of chemical species (B). Figure S8: ^{1}H-NMR spectra of WSF crude oil before (A) and after (B) sonolysis. Figure S9: ^{1}H-NMR compositional analysis of WSF crude oil before (red column) and after (blue column) sonolysis: distribution of the compounds as a function of chemical species. Figure S10: Fluorescence spectra of WSF crude oil before (blu line) and after (red line) sonolysis. Figure S11: GC-MS compositional analysis of WSF crude oil before (red column) and after (blue column) sonophotocatalysis: distribution of hydrocarbons as a function of the number of carbon atoms (A) and distribution of the compounds as a function of chemical species (B). Figure S12: ^{1}H-NMR spectra of WSF crude oil before (A) and after (B) sonophotocatalysis. Figure S13: ^{1}H-NMR compositional analysis of WSF crude oil before (red column) and after (blue column) sonophotocatalysis: distribution of the compounds as a function of chemical species. Figure S14: Fluorescence spectra of WSF crude oil before (blu line) and after (red line) sonophotocatalysis.

Author Contributions: Conceptualisation, F.L., S.A.B. and L.S.; Data curation, F.L., G.B. and S.A.B.; Formal analysis, L.S.; Investigation, L.S.; Methodology, G.B.; Resources, S.A.B.; Writing—original draft, F.L. and L.S.; Writing—review and editing, S.A.B. All authors have read and agreed to the published version of the manuscript.

Funding: This research received no external funding.

Data Availability Statement: Data supporting results reported in this paper can be found at the Department of Sciences, University of Basilicata, Via dell'Ateneo Lucano 10, 85100, Potenza, Italy. Due to the multitude of files produced for this investigation, a link will be provided under specific requests.

Acknowledgments: We acknowledge the Director of ENI-COVA Oil Plant in Val d'Agri (Basilicata Region, Southern Italy), who kindly provided the oil sample. Special thanks go to Mauro Tummolo, and to Jean-Marc Chovelon, University Claud Bernard Lyon 1, for the sonolysis and sonophotocatalysis data.

Conflicts of Interest: The authors declare no conflict of interest.

References

1. Richard Parent. Available online: https://www.experts.com/articles/toxicity-of-crude-oil-and-its-vapors-by-richard-parent (accessed on 17 April 2020).
2. McKee, R.H.; White, R. The Mammalian Toxicological Hazards of Petroleum-Derived Substances: An Overview of the Petroleum Industry Response to the High Production Volume Challenge Program. *Int. J. Toxicol.* **2014**, *33* (Suppl. 1), 4S–16S. [CrossRef]
3. Laffon, B.; Pásaro, E.; Valdiglesias, V. Effects of exposure to oil spills on human health: Updated review. *J. Toxicol. Environ. Health Part B* **2016**, *19*, 105–128. [CrossRef] [PubMed]
4. Guarino, C.; Spada, V.; Sciarrillo, R. Assessment of three approaches of bioremediation (Natural Attenuation, Landfarming and Bioagumentation–Assistited Landfarming) for a petroleum hydrocarbons contaminated soil. *Chemosphere* **2017**, *170*, 10–16. [CrossRef]
5. Cheng, Y.; Wang, L.; Faustorilla, V.; Megharaj, M.; Naidu, R.; Chen, Z. Integrated electrochemical treatment systems for facilitating the bioremediation of oil spill contaminated soil. *Chemosphere* **2017**, *175*, 294–299. [CrossRef] [PubMed]
6. Safdari, M.-S.; Kariminia, H.-R.; Rahmati, M.; Fazlollahi, F.; Polasko, A.; Mahendra, S.; Wilding, W.V.; Fletcher, T.H. Development of bioreactors for comparative study of natural attenuation, biostimulation, and bioaugmentation of petroleum-hydrocarbon contaminated soil. *J. Hazard. Mater.* **2018**, *342*, 270–278. [CrossRef] [PubMed]
7. Bao, M.T.; Wang, L.N.; Sun, P.Y.; Cao, L.X.; Zou, J.; Li, Y.M. Biodegradation of crude oil using an efficient microbial consortium in a simulated marine environment. *Mar. Pollut. Bull.* **2012**, *64*, 1177–1185. [CrossRef]
8. Adzigbli, L.; Bacosa, H.P.; Deng, Y. Response of microbial communities to oil spill in the Gulf of Mexico: A review. *Afr. J. Microbiol. Res.* **2018**, *12*, 536–545. [CrossRef]
9. Al-Gharabally, D.; Al-Barood, A. Kuwait environmental remediation program (KERP): Remediation demonstration strategy. *Biol. Chem. Res.* **2015**, *2015*, 289–296. Available online: http://www.ss-pub.org/wp-content/uploads/2015/09/BCR2015051501.pdf. (accessed on 12 September 2020).
10. Shankar, R.; Won, J.-S.; An, J.-G.; Yim, U.-H. A practical review on photooxidation of crude oil: Laboratory lamp setup and factors affecting it. *Water Res.* **2015**, *68*, 304–315. [CrossRef]
11. Ray, P.Z.; Chen, H.; Podgorski, D.C.; McKenna, A.M.; Tarr, M.A. Sunlight creates oxygenated species in water-soluble fractions of deep water horizon oil. *J. Hazard. Mater.* **2014**, *280*, 636–643. [CrossRef]
12. Vaughan, P.P.; Wilson, T.; Kamerman, R.; Hagy, M.E.; McKenna, A.; Chen, H.; Jeffrey, W.H. Photochemical changes in water accommodated fractions of MC252 and surrogate oil created during solar exposure as determined by FT-ICR MS. *Mar. Pollut. Bull.* **2016**, *104*, 262–268. [CrossRef]
13. Lee, K.; Boufadel, M.; Chen, B.; Foght, J.; Hodson, P.; Swanson, S.; Venosa, A. *Expert Panel Report on the Behaviour and Environmental Impacts of Crude Oil Released into Aqueous Environments*; Royal Society of Canada: Ottawa, ON, Canada, 2015; pp. 38–415. ISBN 978-1-928140-02-3. Available online: https://www.rsc-src.ca/sites/default/files/OIW%20Report_1.pdf. (accessed on 12 September 2020).
14. Cho, E.; Park, M.; Hur, M.; Kang, G.; Kim, Y.H.; Kim, S. Molecular-level investigation of soils contaminated by oil spilled during the Gulf War. *J. Hazard. Mater.* **2019**, *373*, 271–277. [CrossRef] [PubMed]
15. Islam, A.; Cho, Y.; Yim, U.H.; Shim, W.J.; Kim, Y.H.; Kim, S. The comparison of naturally weathered oil and artificially photo-degraded oil at the molecular level by a combination of SARA fractionation and FT-ICR MS. *J. Hazard. Mater.* **2013**, *263*, 404–411. [CrossRef]
16. Lelario, F.; Brienza, M.; Bufo, S.A.; Scrano, L. Effectiveness of different advanced oxidation processes (AOPs) on the abatement of the model compound mepanipyrim in water. *J. Photochem. Photobiol. A Chem.* **2016**, *321*, 187–201. [CrossRef]
17. Hall, G.J.; Frysinger, G.S.; Aeppli, C.; Carmichael, C.A.; Gros, J.; Lemkau, K.L.; Nelson, R.K.; Reddy, C.M. Oxygenated weathering products of Deepwater Horizon oil come from surprising precursors. *Mar. Pollut. Bull.* **2013**, *75*, 140–149. [CrossRef] [PubMed]
18. Prince, R.C.; McFarlin, K.M.; Butler, J.D.; Febbo, E.J.; Wang, F.C.; Nedwed, T.J. The primary biodegradation of dispersed crude oil in the sea. *Chemosphere* **2013**, *90*, 521–526. [CrossRef] [PubMed]
19. Gros, J.; Reddy, C.M.; Aeppli, C.; Nelson, R.K.; Carmichael, C.A.; Arey, J.S. Resolving biodegradation patterns of persistent saturated hydrocarbons in weathered oil samples from the Deepwater Horizon disaster. *Environ. Sci. Technol.* **2014**, *48*, 1628–1637. [CrossRef]
20. Seidel, M.; Kleindienst, S.; Dittmar, T.; Joye, S.B.; Medeiros, P.M. Biodegradation of crude oil and dispersants in deep seawater from the Gulf of Mexico: Insights from ultra-high resolution mass spectrometry. *Deep Sea Res. Part II Top. Stud. Oceanogr.* **2016**, *129*, 108–118. [CrossRef]
21. Bacosa, H.P.; Erdner, D.L.; Liu, Z. Differentiating the roles of photooxidation and biodegradation in the weathering of Light Louisiana Sweet crude oil in surface water from the Deepwater Horizon site. *Mar. Pollut. Bull.* **2015**, *95*, 265–272. [CrossRef] [PubMed]
22. Barrow, M.P.; Peru, K.M.; Fahlman, B.; Hewitt, L.M.; Frank, R.A.; Headley, J.V. Beyond naphthenic acids: Environmental screening of water from natural sources and the Athabasca oil sands industry using atmospheric pressure photoionisation Fourier transform ion cyclotron resonance mass spectrometry. *J. Am. Soc. Mass Spectrom.* **2015**, *26*, 1508–1521. [CrossRef]
23. D'Auria, M.; Racioppi, R.; Velluzzi, V. Photodegradation of crude oil: Liquid injection and headspace solid-phase microextraction for crude oil analysis by gas chromatography with mass spectrometer detector. *J. Chromatogr. Sci.* **2008**, *46*, 339–344. [CrossRef]

24. D'Auria, M.; Emanuele, L.; Racioppi, R.; Velluzzi, V. Photochemical degradation of crude oil: Comparison between direct irradiation, photocatalysis, and photocatalysis on zeolite. *J. Hazard. Mater.* **2009**, *164*, 32–38. [CrossRef]
25. Kim, D.; Ha, S.Y.; An, J.G.; Cha, S.; Yim, U.H.; Kim, S. Estimating degree of degradation of spilled oils based on relative abundance of aromatic compounds observed by paper spray ionisation mass spectrometry. *J. Hazard. Mater.* **2018**, *359*, 421–428. [CrossRef] [PubMed]
26. Christopher, M.R.; Quinn, J.G. GC-MS analysis of total petroleum hydrocarbons and polycyclic aromatic hydrocarbons in seawater samples after the North Cape oil spill. *Mar. Pollut. Bull.* **1999**, *38*, 126–135. [CrossRef]
27. de Oteyza, T.G.; Grimalt, J.O. GC and GC–MS characterisation of crude oil transformation in sediments and microbial mat samples after the 1991 oil spill in the Saudi Arabian Gulf coast. *Environ. Pollut.* **2006**, *139*, 523–531. [CrossRef] [PubMed]
28. Gonzalez, J.J.; Vinas, L.; Franco, M.A.; Fumega, J.; Soriano, J.A.; Grueiro, G.; Muniategui, S.; Lopez-Mahia, P.; Prada, D.; Bayona, J.M.; et al. Spatial and temporal distribution of dissolved/dispersed aromatic hydrocarbons in seawater in the area affected by the Prestige oil spill. *Mar. Pollut. Bull.* **2006**, *53*, 250–259. [CrossRef]
29. McKenna, A.M.; Nelson, R.K.; Reddy, C.M.; Savory, J.J.; Kaiser, N.K.; Fitzsimmons, J.E.; Marshall, A.G.; Rodgers, R.P. Expansion of the analytical window for oil spill characterisation by ultrahigh resolution mass spectrometry: Beyond gas chromatography. *Environ. Sci. Technol.* **2013**, *47*, 7530–7539. [CrossRef] [PubMed]
30. Haleyur, N.; Shahsavari, E.; Mansur, A.A.; Koshlaf, E.; Morrison, P.D.; Osborn, A.M.; Ball, A.S. Comparison of rapid solvent extraction systems for the GC–MS/MS characterisation of polycyclic aromatic hydrocarbons in aged, contaminated soil. *MethodsX* **2016**, *3*, 364–370. [CrossRef] [PubMed]
31. Oldenburg, T.B.; Jones, M.; Huang, H.; Bennett, B.; Shafiee, N.S.; Head, I.; Larter, S.R. The controls on the composition of biodegraded oils in the deep subsurface—Part 4. Destruction and production of high molecular weight non-hydrocarbon species and destruction of aromatic hydrocarbons during progressive in-reservoir biodegradation. *Org. Geochem.* **2017**, *114*, 57–80. [CrossRef]
32. Griffiths, M.T.; Da Campo, R.; O'Connor, P.B.; Barrow, M.P. Throwing light on petroleum: Simulated exposure of crude oil to sunlight and characterisation using atmospheric pressure photoionisation Fourier transform ion cyclotron resonance mass spectrometry. *Anal. Chem.* **2013**, *86*, 527–534. [CrossRef]
33. Lemkau, K.L.; McKenna, A.M.; Podgorski, D.C.; Rodgers, R.P.; Reddy, C.M. Molecular evidence of heavy-oil weathering following the M/V Cosco Busan spill: Insights from Fourier transform ion cyclotron resonance mass spectrometry. *Environ. Sci. Technol.* **2014**, *48*, 3760–3767. [CrossRef]
34. Marshall, A.G.; Rodgers, R.P. Petroleomics: The Next Grand Challenge for Chemical Analysis. *Acc. Chem. Res.* **2004**, *37*, 53–59. [CrossRef]
35. Kendrick, E. A Mass Scale Based on $CH_2=14.0000$ for High Resolution Mass Spectrometry of Organic Compounds. *Anal. Chem.* **1963**, *35*, 2146–2154. [CrossRef]
36. Hughey, C.A.; Hendrickson, C.L.; Rodgers, R.P.; Marshall, A.G.; Qian, K. Kendrick mass defect spectroscopy: A compact visual analysis for ultrahigh-resolution broadband mass spectra. *Anal. Chem.* **2001**, *73*, 4676–4681. [CrossRef] [PubMed]
37. Hughey, C.A.; Rodgers, R.P.; Marshall, A.G. Resolution of 11,000 compositionally distinct components in a single electrospray ionisation Fourier transform ion cyclotron resonance mass spectrum of crude oil. *Anal. Chem.* **2002**, *74*, 4145–4149. [CrossRef]
38. van Krevelen, D.W. Graphical-statistical method for the study of structure and reaction processes of coal. *Fuel* **1950**, *29*, 269–284.
39. Kim, S.; Kramer, R.W.; Hatcher, P.G. Graphical method for analysis of ultrahigh-resolution broadband mass spectra of natural organic matter, the van Krevelen diagram. *Anal. Chem.* **2003**, *75*, 5336–5344. [CrossRef]
40. Hughey, C.A.; Rodgers, R.P.; Marshall, A.G.; Qian, K.; Robbins, W.R. Identification of acidic NSO compounds in crude oils of different geochemical origins by negative ion electrospray Fourier transform ion cyclotron resonance mass spectrometry. *Org. Geochem.* **2002**, *33*, 743–759. [CrossRef]
41. Wu, Z.; Jernström, S.; Hughey, C.A.; Rodgers, R.P.; Marshall, A.G. Resolution of 10,000 compositionally distinct components in polar coal extractsby negative-ion electrospray ionisation Fourier transform ion cyclotron resonance mass spectrometry. *Energy Fuels* **2003**, *17*, 946–953. [CrossRef]
42. Finch, B.E.; Stefansson, E.S.; Langdon, C.J.; Pargee, S.M.; Blunt, S.M.; Gage, S.J.; Stubblefield, W.A. Photo-enhanced toxicity of two weathered Macondo crude oils to early life stages of the eastern oyster (Crassostrea virginica). *Mar. Pollut. Bull.* **2016**, *113*, 316–323. [CrossRef] [PubMed]
43. Cho, Y.; Ahmed, A.; Islam, A.; Kim, S. Developments in FT-ICR MS instrumentation, ionisation techniques, and data interpretation methods for petroleomics. *Mass Spectrom. Rev.* **2015**, *34*, 248–263. [CrossRef] [PubMed]
44. Pradyot, P. *A Comprehensive Guide to the Hazardous Properties of Chemical Substances*; Wiley & Sons: New York, NY, USA, 2007; pp. 161–568.
45. Khalaf, S.; Shoqeir, J.H.; Lelario, F.; Bufo, S.A.; Karaman, R.; Scrano, L. TiO_2 and Active Coated Glass Photodegradation of Ibuprofen. *Catalysts* **2020**, *10*, 560. [CrossRef]

Article

Visible Light Responsive Strontium Carbonate Catalyst Derived from Solvothermal Synthesis

Pornnaphat Wichannananon [1,2], Thawanrat Kobkeatthawin [2] and Siwaporn Meejoo Smith [2,*]

[1] Center of Excellence for Innovation in Chemistry, Faculty of Science, Mahidol University, 272 Rama VI Rd., Rajthevi, Bangkok 10400, Thailand; pornnaphat.w@gmail.com

[2] Center of Sustainable Energy and Green Materials and Department of Chemistry, Faculty of Science, Mahidol University, 999 Phuttamonthon Sai 4 Road, Salaya, Nakorn Pathom 73170, Thailand; kunthidakob@gmail.com

* Correspondence: siwaporn.smi@mahidol.edu; Tel.: +66-93593-9449

Received: 24 July 2020; Accepted: 11 September 2020; Published: 17 September 2020

Abstract: A single crystalline phase of strontium carbonate ($SrCO_3$) was successfully obtained from solvothermal treatments of hydrated strontium hydroxide in ethanol (EtOH) at 100 °C for 2 h, using specific Sr:EtOH mole ratios of 1:18 or 1:23. Other solvothermal treatment times (0.5, 1.0 and 3 h), temperatures (80 and 150 °C) and different Sr:EtOH mole ratios (1:13 and 1:27) led to formation of mixed phases of Sr-containing products, $SrCO_3$ and $Sr(OH)_2$ xH_2O. The obtained products (denoted as 1:18 $SrCO_3$ and 1:23 $SrCO_3$), containing a single phase of $SrCO_3$, were further characterized in comparison with commercial $SrCO_3$, and each $SrCO_3$ material was employed as a photocatalyst for the degradation of methylene blue (MB) in water under visible light irradiation. Only the 1:23 $SrCO_3$ sample is visible light responsive (E_g = 2.62 eV), possibly due to the presence of ethanol in the structure, as detected by thermogravimetric analysis. On the other hand, the band gap of 1:18 $SrCO_3$ and commercial $SrCO_3$ are 4.63 and 3.25 eV, respectively, and both samples are UV responsive. The highest decolourisation efficiency of MB solutions was achieved using the 1:23 $SrCO_3$ catalyst, likely due to its narrow bandgap. The variation in colour removal results in the dark and under visible light irradiation, with radical scavenging tests, suggests that the high decolourisation efficiency was mainly due to a generated hydroxyl-radical-related reaction pathway. Possible degradation products from MB oxidation under visible light illumination in the presence of $SrCO_3$ are aromatic sulfonic acids, dimethylamine and phenol, as implied by MS direct injection measurements. Key findings from this work could give more insight into alternative synthesis routes to tailor the bandgap of $SrCO_3$ materials and possible further development of cocatalysts and composites for environmental applications.

Keywords: strontium carbonate ($SrCO_3$); solvothermal method; photocatalysis; visible light

1. Introduction

Textile industries employ over 10,000 dyes and pigments in the manufacturing of cotton, leather, clothes, wool, silk and nylon products [1–3]. An estimated 700,000 tons or more of synthetic dyes are thought to be annually discharged into the environment [4], causing serious water pollution as many of these dyes are toxic, highly water soluble and highly stable against degradation by sunlight or increased temperature [5]. Therefore, effective treatments of dye-contaminated water have continuously received great attention by academic and industrial sectors. Various wastewater treatment methods have been applied to remove toxic dyes from wastewater, such as coagulation–flocculation, adsorption, membrane separation, biodegradation and oxidation processes [6]. Among these methods, photocatalytic oxidation processes have been proven to be simple and effective at organic dye decomposition, forming relatively low toxic by-products with potential mineralization to generate CO_2 and H_2O [7–9]. In this process, under light irradiation a semiconducting catalyst absorbs photon energy promoting electron transfer from the valence band (VB) to the conduction band (CB), resulting in electron-hole pair generation. The generated holes (h+) further react with water molecules while the electrons (CB) react with oxygen, resulting in formation of active hydroxyl (•OH) and superoxide (•O_2^-) radicals, respectively. The •OH radicals subsequently attack organic pollutants in water leading to oxidative degradation of pollutants.

Wide bandgap TiO_2 [10] and ZnO [11] semiconducting materials have proven to be efficient catalysts for the photo-oxidation of organic pollutants in water. However, these require high-energy ultraviolet irradiation, which requires special and costly safety protocols to be in place for the use of these materials in wastewater treatment. An attractive alternative is to use harmless visible light sources in the photoreactor, employing a visible light responsive photocatalyst for pollutant degradation. Such visible light responsive photocatalysts need to promote the photo-oxidative degradation of pollutants using sunlight (7% UV and 44% visible light emission, and other low-energy radiations [12]) to ensure wastewater treatment is a sustainable process. Strontium carbonate ($SrCO_3$) is a common starting material for the manufacture of colourants in fireworks, glass cathode-ray tubes and computer monitors [13,14]. While commercially available $SrCO_3$ material is commonly derived from celestine ($SrSO_4$) mineral via calcination followed by Na_2CO_3 treatment (the black ash method) [15], synthetic $SrCO_3$ can be obtained using calcination and wet chemical methods under ambient [16,17] or high-pressure [18] atmospheres. Table 1 summarizes the key features (synthesis conditions, characteristics and the bandgap energy) of synthetic $SrCO_3$ in the literature. Methylene blue, a cationic organic dye and a common colouring agent used in cotton, wood, silk [19] cosmetics, and textile [20] dying is a frequently utilized representative dye pollutant mimicking those present in industrial effluents. Song and coworkers reported effective methylene blue (MB) degradation under visible light irradiation ($\lambda > 400$ nm) after 3 h treatment with $SrCO_3$ obtained from the calcination of synthetic $Sr(OH)_2$ [21], while Molduvan and coworkers reported the removal of MB from aqueous solutions using a commercial natural activated plant-based carbon [22]. Other works have utilized $SrCO_3$ as a cocatalyst incorporated in photocatalyst composites, e.g., $Ag_2CO_3/SrCO_3$ [23], $TiO_2/SrCO_3$ [24] and $SrTiO_3/SrCO_3$ [25], to expand the photoresponsive range of the material and to improve its catalytic activity and reaction selectivity.

Table 1. Synthesis, key characteristics and bandgap energy of synthetic $SrCO_3$.

Conditions	Reaction Time	Morphology	Band Gap	References
Celetine ore ($SrSO_4$) (industrial scale)	Reductive calcination followed by Na_2CO_3(aq) assisted precipitation (the black ash method)	-	-	[26]
Celetine ore ($SrSO_4$) (industrial scale)	Double decomposition in Na_2CO_3(aq)	-	-	[27]
$Sr(OH)_2 + CO_2$ + EDTA	50 °C, 10 min	spherical shape	-	[28]
$Sr(NO_3)_2$ + TEA + NaOH	100 °C, 12 h	branchlet-like $SrCO_3$ nanorods	-	[17]
$SrCl_2 + H_2O$ + DMF + glycerol	120 °C, 8 h	flower shape	-	[29]
$Sr(NO_3)_2$ + urea	120 °C, 24 h	urchin-like $SrCO_3$	-	[30]
$Sr(NO_3)_2 + (NH_4)_2CO_3$ + HMT	Room temp, 7 days	branch-like, flower-like, capsicum-like	-	[31]
$CH_3COO_2Sr + Na_2SO_4$ + hexamethylene triamine	160 °C, 24 h	spherical shape	3.17	[32]
$Sr(NO_3)_2 + Na_2CO_3$	120 °C, 8 h	various shapes such as rod shape ellipsoid shape sphere shape	-	[33]
1. $SrCl_2 + Na_2CO_3$ 2. $SrCl_2 + CO(NH_2)_2$ 3. $SrCl_2 + CO(NH_2)_2$ + SDS	1. 110 °C, 12 h 2. 110 °C, 12 h 3. 110 °C, 12 h	rod shape rod shape flower shape	-	[34]
$Sr(OH)_2$ + flowing CO_2 gas	1. 50 °C, 12 h 2. 60 °C, 12 h 3. 70 °C, 12 h	nanowhisker rod shape pherical shape	-	[35]

This work investigated the effects of precursor concentrations (Sr:ethanol mole ratios), solvothermal temperatures and treatment times on the properties of $SrCO_3$ materials and their photocatalytic degradation of MB in water under visible light irradiation, as a function of pH and temperature. Kinetic and mechanistic studies of the MB degradation process were carried out through reaction rate determination and identification of the end-products. The photocatalytic performance of synthesized $SrCO_3$ was compared with that of commercially available material, in order to derive insights into the relationships between properties and catalytic activity.

2. Results and Discussions

2.1. Effects of Synthesis Conditions

Solvothermal treatments of strontium nitrate in ethanol (EtOH) were carried out at various temperatures (80, 100 and 150 °C), treatment times (0.5, 1, 2 and 3 h) and Sr:EtOH mole ratios (1:13, 1:18, 1:23 and 1:27). From powder X-ray diffraction (PXRD) results in Figure 1, a single phase of $SrCO_3$ was obtained from two conditions: 2 h solvothermal treatment at 100 °C using a Sr:EtOH mole ratio of 1:18 or 1:23. These samples are denoted as 1:18 $SrCO_3$ and 1:23 $SrCO_3$ in further discussions. Notably, mixed phases of $SrCO_3$ and hydrated strontium hydroxides ($Sr(OH)_2 \cdot xH_2O$, where x is the number of molar coefficient of water in strontium hydroxide solid) were obtained from all other synthesis conditions (results shown in Supplementary Materials: Figures S1 and S2). Typical diffraction peaks correspond well with (110), (111), (021), (002), (012), (130), (220), (221), (132) and (113) orthorhombic $SrCO_3$ lattice planes [34,36], whereas other diffraction peaks match with those of previously reported $Sr(OH)_2 \cdot H_2O$ [37] and $Sr(OH)_2 \cdot 8H_2O$ phases [38]. The formation of $Sr(OH)_2 \cdot xH_2O$ is possibly due to adsorbed alcohol, promoting the addition of OH functional groups on the solid surface [39], upon solvothermal crystallization of Sr-containing products.

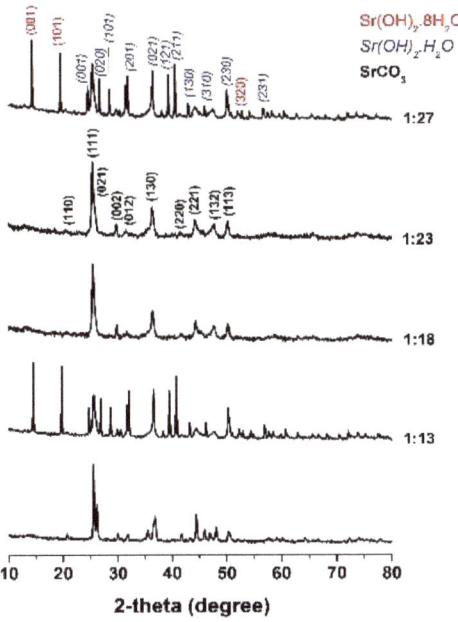

Figure 1. PXRD patterns of Sr-containing samples derived from 2 h solvothermal treatments of hydrated strontium hydroxide in ethanol at various Sr:EtOH mole ratios (1:13, 1:18, 1:23 or 1:27) at 100 °C.

FTIR spectra of the prepared Sr-containing samples are shown in Figure 2. The absorption bands located within 1700–400 cm^{-1} regions were attributed to the vibrations in CO_3^{2-} groups. The strong broad absorption at 1470 cm^{-1} was considered to be due to an asymmetric stretching vibration, and the sharp absorption bands at 800 cm^{-1} and 705 cm^{-1} can be specified to the bending out-of-plane vibration and in-plane vibration, respectively. The weak peak at 1770 cm^{-1} indicated a combination of vibration modes of the CO_3^{2-} groups and Sr^{2+}. The sharp peak at 3500 cm^{-1} was assigned to the stretching mode of –OH- in $Sr(OH)_2$, and the broad absorption peak around 2800 cm^{-1} was assigned to the stretching mode of H_2O in $Sr(OH)_2 \cdot H_2O$ and $Sr(OH)_2 \cdot 8H_2O$. These results are consistent with the commercial $SrCO_3$ and the 1:18 $SrCO_3$ and 1:23 $SrCO_3$ samples being of similar chemical composition.

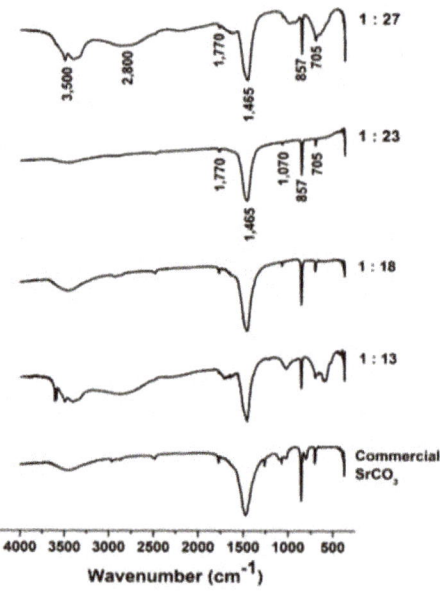

Figure 2. FTIR spectra of Sr-containing samples derived from 2 h solvothermal treatment of hydrated strontium hydroxide in ethanol at various Sr:EtOH mole ratios (1:13, 1:18, 1:23 or 1:27) at 100 °C.

SEM images of the obtained $SrCO_3$ materials (derived from Sr:EtOH mole ratios of 1:18 or 1:23) are compared with those of commercial $SrCO_3$ in Figure 3. Whisker-like $SrCO_3$ and spherical particles were obtained under these respective synthesis conditions. Figure 3c highlights the relatively large rod-like particles of commercial $SrCO_3$. Variation in particle sizes was observed in solvothermally obtained $SrCO_3$, with particle sizes being smaller for the 1:18 $SrCO_3$ samples. Notably, commercial $SrCO_3$ contains much larger particles than those of the synthesized material. From literature [26,27], $SrCO_3$ production plants utilize two common methods, the black ash method and the soda method, in conversion of celestine ore ($SrSO_4$) to $SrCO_3$ (Table 1). The black ash method involves high-temperature calcination of the ore to obtain SrS, with crystalline $SrCO_3$ solid being formed after dissolving the SrS in aqueous Na_2CO_3, followed by precipitation. The soda method produces $SrCO_3$ through the two-step decomposition reaction between celestine and aqueous Na_2CO_3, to obtain precipitated $SrCO_3$. From this information, as the formation of commercial $SrCO_3$ does not require high temperatures (>150 °C) for solvent evaporation and precipitation of $SrCO_3$, the larger grain size of the commercial $SrCO_3$ sample is probably due to the fast solvent evaporation during the precipitation processes.

Figure 3. SEM images of SrCO$_3$ derived from 2 h solvothermal treatment at 100 °C, using Sr:EtOH mole ratios of (**a**) 1:18 and (**b**) 1:23 compared with (**c**) commercial SrCO$_3$.

Thermogravimetric analysis (TGA) plots (Figure 4) suggest thermal stability of all SrCO$_3$ samples up to 600 °C. Slight weight loss (<1%) was likely due to moisture or solvent residue [40]. The 1:23 SrCO$_3$ sample gives a relatively high weight loss of 0.21%, which corresponds to the removal of surface adsorbed moisture and ethanol (weight loss upon heating up to 400 °C) and the loss of ethanol from the SrCO$_3$ lattice at ca. 450 °C. Decomposition of SrCO$_3$ takes place at temperatures above 800 °C as a result of conversion to SrO.

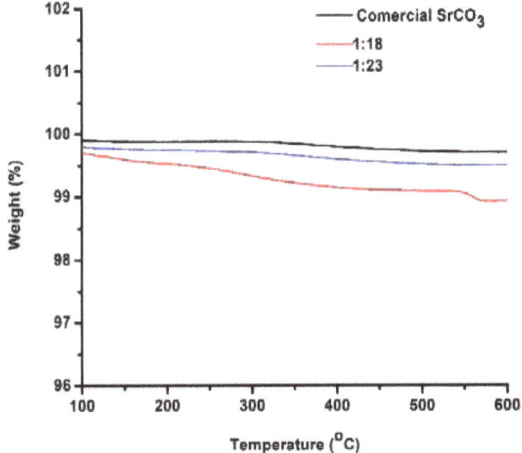

Figure 4. Thermogravimetric analysis (TGA) plots of SrCO$_3$ samples prepared using Sr:EtOH mole ratios of 1:18 and 1:23, compared with that of commercial SrCO$_3$.

Based on the PXRD and TGA results, chemical transformation of hydrated strontium hydroxide in the presence of ethanol under solvothermal treatments leads to the formation of SrCO$_3$ and ethanol incorporated SrCO$_3$ materials, as proposed by the reactions below. In general, CO$_2$ in air can react with strontium hydroxide to form SrCO$_3$, which precipitates after the sonication step and solvothermal treatments. Ethoxide could be formed under basic conditions, resulting in an CH$_3$CH$_2$O···Sr^{2+}···OCH$_2$CH$_3$ intermediate, which is subsequently transformed to ethanol incorporated in SrCO$_3$. Note that the amount of ethanol incorporated within the SrCO$_3$ is sufficiently low, such that a single phase of SrCO$_3$ was observed in PXRD pattern of the 1:23 SrCO$_3$ sample.

$$Sr(OH)_2 + CO_2 \rightarrow SrCO_3 + H_2O$$

$$Sr(OH)_2 + CO_2 \rightleftharpoons Sr^{2+} + HCO_3^- + OH^-$$

$$HCO_3^- + OH^- \rightarrow CO_3^{2-} + H_2O$$

$$CH_3CH_2OH + OH^- \rightleftharpoons CH_3CH_2O^- + H_2O$$

$$Sr(OH)_2 + H_2O \rightleftharpoons Sr^{2+}\,(aq) + OH^-\,(aq)$$

$$Sr^{2+} + 2CH_3CH_2O^- \rightarrow Sr^{2+}\cdots 2OCH_2CH_3$$

$$Sr^{2+}\cdots OCH_2CH_3 + HCO_3^- \rightarrow SrCO_3 \cdots HOCH_2CH_3$$

2.2. Optical Properties

UV–VIS diffuse reflectance spectra of the 1:18 $SrCO_3$, 1:23 $SrCO_3$ and commercial $SrCO_3$ in Figure 5a showed that the characteristic absorption edge of the 1:23 $SrCO_3$ sample is located in the visible light region (473 nm), whereas the spectral response of other $SrCO_3$ samples was observed in the UV region, with absorption band edges of 268 and 381 nm for the 1:18 $SrCO_3$ sample and commercial $SrCO_3$, respectively. The band gap energy values suggested by Kubelka–Munk plots (Figure 5b) are 4.63 and 3.25 eV for 1:18 $SrCO_3$ and commercial $SrCO_3$, respectively. By contrast, the bandgap energy of the 1:23 $SrCO_3$ sample is 2.62 eV, and its visible response is possibly due to the presence of incorporated ethanol in the solid sample, as suggested by TGA results.

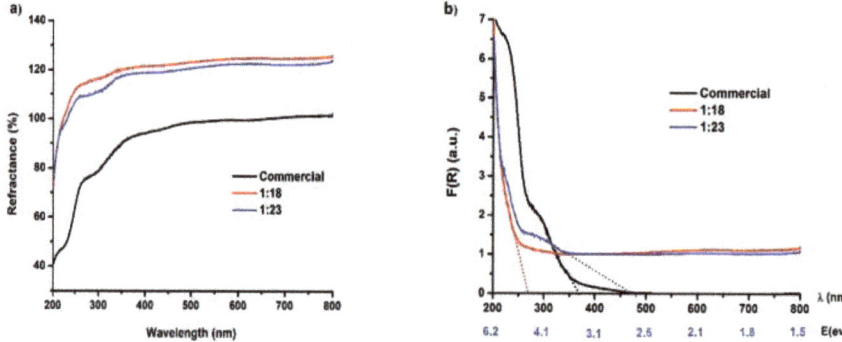

Figure 5. (a) UV-visible diffuse reflectance spectra and (b) Kubelka–Munk plots of the $SrCO_3$ synthesized at Sr:EtOH mole ratios of 1:18 and 1:23, compared with those of commercial $SrCO_3$.

2.3. Decolourisation of Methylene Blue (MB)

Figure 6a illustrates the colour removal efficiencies of 10 ppm MB aqueous solutions in the dark and under visible light irradiation after 1 h treatment with $SrCO_3$. Similar colour removal efficiencies from treatment of MB(aq) with 1:18 $SrCO_3$ in the dark and under light illumination suggested major adsorption processes occurred due to the wide bandgap of the 1:18 $SrCO_3$ sample. On the other hand, the visible responsive 1:23 $SrCO_3$ and commercial $SrCO_3$ gave higher colour removal efficiencies under irradiation conditions than those from dark experiments, implying both adsorption and photodegradation of MB are of importance. Therefore, from these catalyst screening tests, the colour removal efficiencies of aqueous MB solutions strongly depend on the bandgap energy of $SrCO_3$ materials and that the 1:23 $SrCO_3$ is the most active catalyst. Figure 6b demonstrates that only low colour removal efficiencies occur due to adsorption (in the dark) and photolysis (irradiation and no $SrCO_3$). Treatments of dye solutions with 1:23 $SrCO_3$ is much less effective (low colour removal efficiency) under dark conditions in comparison to decolourisation under visible light irradiation. These results suggest that the main process of MB colour removal is caused by photocatalytic treatment by using the $SrCO_3$ photocatalyst rather than adsorption.

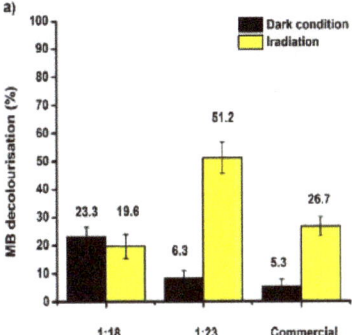
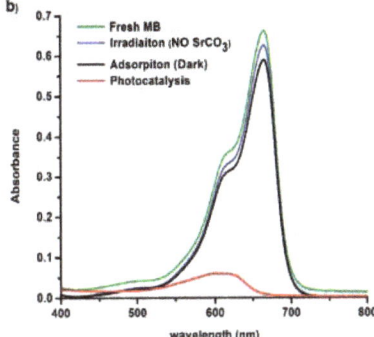

Figure 6. (**a**) Colour removal efficiencies of 10 ppm methylene blue (MB) aqueous solution in the dark, and under visible light irradiation in the presence of SrCO$_3$. (**b**) Absorption spectra of MB in the dark or under visible light irradiation by SrCO$_3$ (1:23 and 1:18). All decolourisation experiments were performed at 30 °C using SrCO$_3$ with catalyst loadings of 4.0 g·L^{-1} with 1 h treatment.

The percentage of MB colour removal after treatment with SrCO$_3$ photocatalyst (sample 1:23) is shown in Figure 7a. When a suspension of SrCO$_3$ in 10 ppm fresh MB solution was kept in the dark for 3 h, the concentration of dye slightly decreased, while the colour of the dye solution remained unchanged. It was observed that the absorption capacity of MB on the SrCO$_3$ surface is negligible because the specific area of the prepared SrCO$_3$ photocatalyst is low (9.23 m^2·g^{-1}). Upon visible irradiation, the prepared SrCO$_3$ gave a high percentage of MB colour removal (>99% after 3 h visible irradiation).

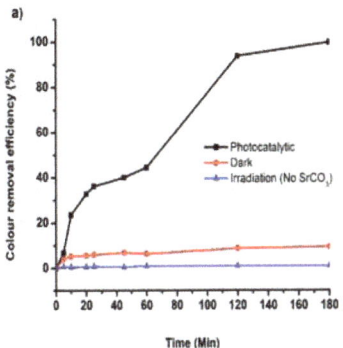
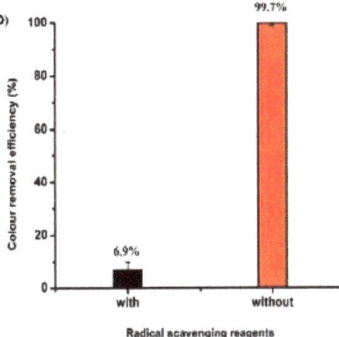

Figure 7. (**a**) Colour removal efficiencies of 10 ppm MB aqueous solution (pH 5.5) as a function of time in the dark or under visible light irradiation; adsorption of MB into SrCO$_3$ (loading 4 g·L^{-1}) in the dark, MB photolysis and photocatalysis of MB under visible light illumination catalyzed by SrCO$_3$ (loading 4 g·L^{-1}). (**b**) Effects of a scavenger (tert-BuOH) on the colour removal efficiency of 10 ppm MB after 3 h treatment with the 1:23 SrCO$_3$ (4 g·L^{-1}).

In order to prove that hydroxyl radicals (•OH) are the active species in the photocatalytic degradation process, experiments were conducted in the presence of a radical scavenging reagent. One such reagent, tert-butyl alcohol (tert-BuOH), if present, should significantly inhibit the oxidation of MB [41]. The result in Figure 7b indicates that after treatment for 3 h, adding tert-BuOH resulted in poor colour removal efficiencies (6.90%), whereas in the absence of the reagent very high colour removal efficiencies (>99%) were achieved. The formation of a product arising from the reaction between tert-BuOH and •OH as ascribed through a radical pathway [41] thus resulted in the poor activity, confirming that hydroxyl radicals are the important active species assisting MB degradation.

The effect of pH on the MB decolourisation under visible light irradiation was examined over a range of pH 3–9. The colour removal efficiency reached 73% after 1 h treatment at pH 3, while lower colour removal efficiencies were obtained at pH 5.5 (51%), pH 7 (42%) and pH 9 (29%) over the same time period, as shown in Figure 8a. In addition, the natural logarithm of the MB concentrations was plotted as a function of irradiation time, affording a linear relationship, as presented in Figure 8b. Using the first-order model, the highest rate constant of MB colour removal was obtained at pH 3, with the degradation being slowest at pH 9. The decreasing rate constants of MB decolourisation with increasing pH may be the result of the presence of carbonate (CO_3^{2-}) and hydroxide (OH^-) ions, which are radical scavengers [42,43]. At pH 5.5–10, the low colour removal efficiencies may be due to the following reactions.

$$CO_3^{2-} + \bullet OH \rightarrow CO_3^{\bullet -} + OH^-$$

$$OH^- + \bullet OH \rightarrow H_2O + O^-$$

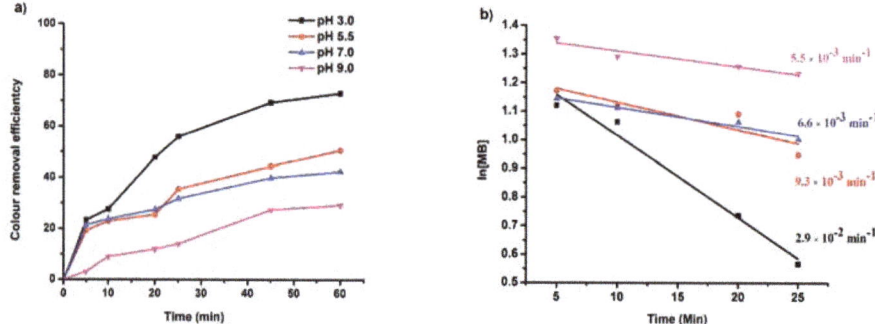

Figure 8. (a) Colour removal efficiencies of 10 ppm MB aqueous solution with time, using $SrCO_3$ as photocatalyst. (b) Kinetics of MB decolourisation catalyzed by $SrCO_3$ as a function of pH. All decolourisation experiments were performed using $SrCO_3$ with catalyst loading of 4.0 g·L^{-1} at 30 °C from pH 3–9.

The effect of temperature on the degradation of MB as a function of time is discussed in Figure 9. From Figure 9a, it can be observed that higher temperatures result in higher MB colour removal efficiencies. Under visible light irradiation, the MB colour removal efficiency reached 100% after 1 h treatment at 70 °C. In all cases MB concentrations decrease with irradiation time. The linear plots between the natural logarithm of the MB concentration versus irradiation time are shown in Figure 9b, which indicate that the decolourisation process follows first-order kinetics. The rate constants of MB decolourisation increased with temperature, indicating that MB removal by 1:23 $SrCO_3$ is overall endothermic. The 1:23 $SrCO_3$ sample is rather stable during the photocatalytic MB degradation reaction, as only negligible concentrations of Sr (<10 ppm) were detected in the treated MB solution.

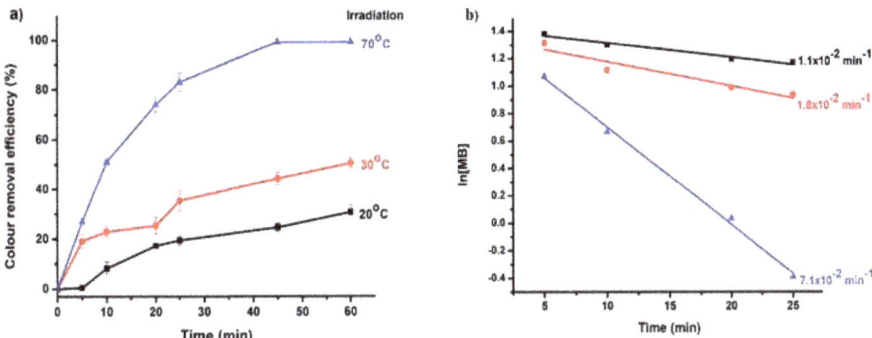

Figure 9. (a) Colour removal efficiencies of 10 ppm MB aqueous solution (pH 5.5) over time using SrCO$_3$ as photocatalyst. (b) Kinetics of MB decolourisation catalyzed by SrCO$_3$ as a function of temperature. All decolourisation experiments were performed using 1:23 SrCO$_3$ with catalyst loading of 4.0 g·L^{-1} at temperatures 20–70 °C.

2.4. Degradation Products

Figure 10 highlights mass spectra generated from the MB degradation products with the mass-to-charge ratios (m/z) of 77, 122, 234, 284 and 303, reported with the possible fragmented ions shown accordingly.

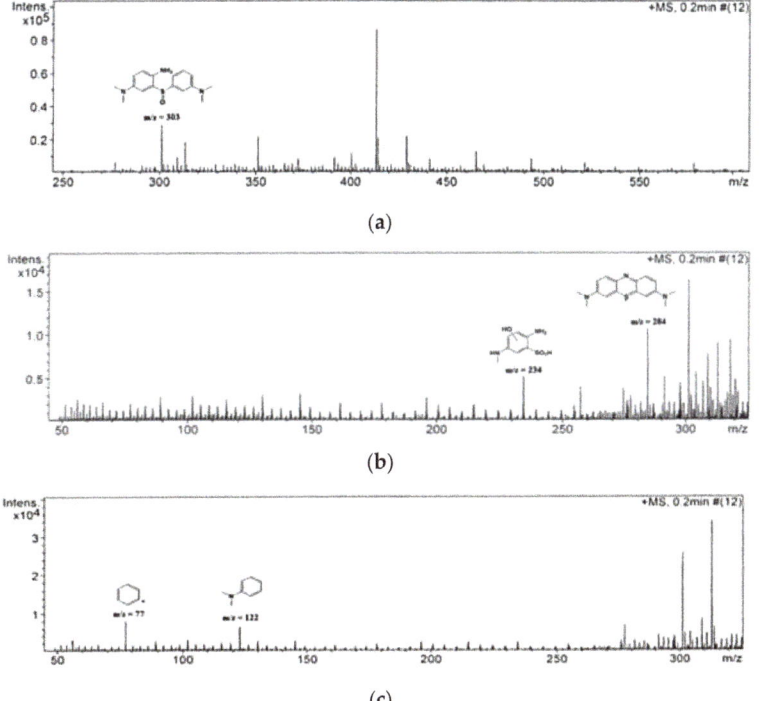

Figure 10. Mass spectra of intermediates from the MB degradation after treatment for (**a**) 10 min, (**b**) 25 min and (**c**) 60 min. All experiments were performed by suspending 1:23 SrCO$_3$ in 10 ppm MB (4 g·L^{-1} of MB solution) followed by visible light illumination.

The proposed reaction pathway of MB photooxidation over SrCO$_3$ photocatalyst is outlined in Figure 11. The detected degradation products, as identified from fragments based on m/z ratio, are illustrated in blue, while undetectable but expected intermediates [44,45] are presented in black. These results are in general agreement with previous works that report the generated intermediates during the MB photodegradation process [44,45].

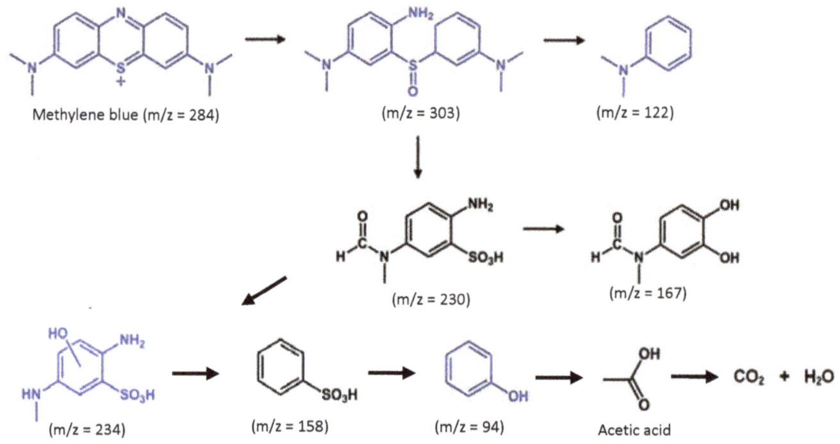

Figure 11. Proposed photocatalytic degradation pathway of MB. Detected degradation products are illustrated in blue, while expected but undetectable [44] species are presented in black.

3. Materials and Methods

3.1. Chemicals

All reagents were used without further purification. Chemicals of HPLC grade were acetic acid (C$_2$H$_3$O$_2$, Merck, Darmstadt, Germany) and acetonitrile (C$_2$H$_2$N, J.T. Baker, CA, USA). Chemicals of AR grade were ethanol (C$_2$H$_5$OH, Merck, Germany), potassium bromide (KBr, Merck, Germany), tert-butanol (C$_4$H$_{10}$O, Merck, Germany), ammonium acetate (C$_2$H$_2$ONH$_4$, Rankem, Gurugram, India), sodium hydroxide (NaOH, Rankem, India), methylene blue (C$_{16}$H$_{18}$N$_3$Cl, Fluka, Saint Louis, MO, USA), strontium carbonate (SrCO$_3$, Fluka, USA), concentrated hydrochloric acid (HCl, Lab Scan, Ireland), mercury(II) sulphate (HgSO$_4$, QRëc, Newzaland), nitric acid (HNO$_3$, Mallinckrodt Chemicals, Phillipsburg, NJ, USA), potassium dichromate (K$_2$Cr$_2$O$_7$, Unilab, Mandaluyong, Philippines), potassium hydrogen phthalate (C$_8$H$_5$KO$_4$, Univar, Redmond to Downers Grove, IL, USA), silver sulphate (AgSO$_4$, Carlo Erba, Barcelona, Spain), strontium hydroxide octahydrate (Sr(OH)$_2$·8H$_2$O, Sigma Aldrich, Saint Louis, MO, USA) and concentrated sulfuric acid (H$_2$SO$_4$, Lab supplies, Spain).

3.2. Synthesis of Strontium Carbonate (SrCO$_3$)

Strontium carbonate (SrCO$_3$) was synthesized by a solvothermal method modified from the procedure of Zhang et al. [34]. A suspension of 20 g Sr(OH)$_2$·8H$_2$O in ethanol (100 mL) was sonicated in an ultrasonic bath for 20 min, followed by solvothermal treatment in an autoclave at 80, 100, 120 or 150 °C for 2 h. The reaction mixtures were left at room temperature to cool down to room temperature. Then, the precipitates were washed with deionized water to remove Sr(OH)$_2$·xH$_2$O, dried and kept in a dry condition at room temperature. After obtaining the optimal treatment temperature, the reaction time was investigated through the above procedure by fixing the treatment temperature at 100 °C and varying reaction time between 0.5, 1, 2 or 3 h. The strontium-based samples were prepared by varying the Sr(OH)$_2$·8H$_2$O: ethanol mole ratio as either 1:13, 1:18, 1:23 or 1:27, and then the above procedures were followed using a treatment temperature of 100 °C for 2 h.

3.3. Materials Characterisation

The crystallinity and the phase structure of the samples were investigated using X-ray diffractometry (PXRD, Bruker AXS model D8 advance). The measurements were examined with CuKα radiation between 2θ values of 10–80 degrees, at a scan rate of 0.075 degree·min^{-1} using accelerating voltage and currents of 40 kV and 40 mA, respectively. Chemical composition and bonding information were probed using Fourier transform infrared spectrophotometry (FT-IR, Elmer model lamda 800). Diffusion reflectance spectra were measured on a UV–VIS spectrophotometer (Agilent Cary 5000) using a scanning rate of 200–1100 nm. Sample morphologies were investigated using scanning electron microscopy (SEM). The thermal decomposition of $SrCO_3$ was monitored using a thermogravimetric analyzer (TGA, TA instruments SDT 2960 Simultaneous DSC-TGA).

3.4. Catalyst Performance Examinations

$SrCO_3$ samples were dispersed in 10 mL of 10 ppm MB aqueous solution in order to observe the change in colour under dark and visible light irradiation conditions. Before illumination, the suspensions were stirred in the dark for 5 min. Then, suspensions were irradiated using an LED (16 × 12 V EnduraLED 10 W MR16 dimmable 4000 K with λ > 400 nm) [46]. The colour removal efficiency of MB was monitored as a function of degradation time by measuring the absorbance of the dye solution after treatment. In order to terminate the reaction, the photocatalyst was filtered off using a syringe filter (0.45 μm). The absorbance of the dye was then measured, and the concentration of remaining MB was quantified using the absorbance at maximum wavelength (around 664.5 nm) using the Beer Lambert law.

The colour removal efficiency of MB was calculated via Equation (1):

$$\text{Colour removal efficiency} = \left(\frac{C_0 - C_t}{C_0}\right) \times 10 \quad (1)$$

where C_0 is the concentration of fresh MB solution, and C_t is the concentration of dye residue after treatment at t minutes.

Leaching of strontium ions may be a major cause of photocatalyst deactivation. Therefore, the amount of strontium ions in the filtered MB solution was quantified by flame atomic absorption spectrometry (FAAS, Perkin Elmer, Waltham, MA, USA).

A mass spectrometer (micro TOF MS, Bruker, Billerica, MA, USA) equipped with electrospray ionization (ESI) source was employed to detect MB degradation products. For this, direct injection of the treated MB solution (with 1:23 $SrCO_3$) under visible light irradiation was carried out, with fragments examined over the range m/z 50–700.

4. Conclusions

In this work, a solvothermal method without any calcination step was employed to prepare a single crystalline phase of strontium carbonate ($SrCO_3$). Ethanol incorporated $SrCO_3$, a visible light responsive $SrCO_3$ material having a bandgap energy of 2.62 eV, was obtained from the solvothermal treatment of hydrated strontium hydroxide in ethanol at Sr:EtOH of 1:23. Nevertheless, the synthesis conditions strongly influence the bandgap energy of $SrCO_3$, as UV responsive $SrCO_3$ material can also be obtained by varying the precursor concentration. The narrow bandgap $SrCO_3$ material can be utilized as a photocatalyst for decolourisation of methylene blue in water under visible light irradiation. Effective decolourisation of 10 ppm methylene blue aqueous solutions was achieved with >99% colour removal efficiencies after 3 h treatment, under visible light irradiation over the 1:23 photocatalyst, using a catalyst loading of 4 g·L^{-1}. The decolourisation is mainly due to photocatalytic processes. The rate constant values showed a direct correlation with temperature, but decolourisation was most rapid at low pH. In addition to the conventional uses of $SrCO_3$ in pyrotechnics and frit manufacturing, synthesized $SrCO_3$ materials have their place as semiconductors and cocatalysts employed in energy and environmental applications. The key findings of this work highlight that incorporated ethanol in the SrCO$_3$ structure results in a narrowing of the energy bandgap in SrCO$_3$, with the material being a visible light responsive semiconductor and active photocatalyst in dye degradation. Results from this work may suggest alternative synthesis routes to obtain visible responsive SrCO$_3$ materials, for further development of new composites and cocatalysts in broader applications.

Supplementary Materials: The following are available online at http://www.mdpi.com/2073-4344/10/9/1069/s1. Figure S1: PXRD patterns of Sr-containing samples derived from solvothermal treatments of hydrated strontium hydroxide in ethanol (a) at various solvothermal temperatures, 2 h, Sr:EtOH mole ratios of 1:23 and (b) at various solvothermal treatment times, 100 °C, Sr:EtOH mole ratios of 1:23; Figure S2: FTIR spectra of Sr-containing samples derived from solvothermal treatment of hydrated strontium hydroxide in ethanol (a) at various solvothermal temperatures, 2 h, Sr:EtOH mole ratios of 1:23 and (b) at various solvothermal treatment times, 100 °C, Sr:EtOH mole ratios of 1:23.

Author Contributions: Formal acquisition, investigation and writing—original draft, P.W. writing—review, editing, T.K.; funding acquisition, writing—review, editing and supervision, S.M.S. All authors have read and agreed to the published version of the manuscript.

Funding: M.Sc. Student scholarship (for P.W.) was provided by the Center of Excellence for Innovation in Chemistry (PERCH-CIC). This work was partially supported by the Thailand Research Fund (Grant No. RSA5980027 and IRN62W0005) for T.K. and S.M.S., the National Research Council of Thailand for P.W, and by the CIF, Faculty of Science, Mahidol University.

Acknowledgments: The authors are grateful for partial financial support from the Thailand Research Fund (Grant No. RSA5980027 and IRN62W0005), the National Research Council of Thailand, and the CIF, Faculty of Science, Mahidol University. PP is thankful for an M.Sc. student scholarship from the Center of Excellence for Innovation in Chemistry (PERCH-CIC).

Conflicts of Interest: The authors declare no conflict of interest.

References

1. Vosoughifar, A. Photodegradation of dye in waste water using CaWO$_4$/NiO nanocomposites co-precipitation preparation and characterization. *J. Mater. Sci. Mater. Electron.* **2018**, *29*, 3194–3200. [CrossRef]
2. Chanathaworn, J.; Bunyakan, C.; Wiyaratn, W.; Chungsiriporn, J. Photocatalytic decolurization of basic dye by TiO$_2$ nanoparticle in photoreactor. *Songklanakarin J. Sci. Technol.* **2012**, *34*, 203–210.
3. El-Shishtawy, R.M.; Al-Zahrani, F.A.M.; Afzal, S.M.; Razvi, M.A.N.; Al-Amshany, Z.M.; Bakry, A.H.; Asiri, A.M. Synthesis, linear and nonlinear optical properties of a new dimethine cyanine dye derived from phenothiazine. *RSC Adv.* **2016**, *6*, 91546–91556. [CrossRef]
4. Gupta, V.K.; Mittal, A.; Gajbe, V.; Mittal, J. Adsorption of basic fuchsin using waste materials bottom ash and deoiled soya-as adsorbents. *J. Colloid Interface Sci.* **2008**, *319*, 30–39. [CrossRef]
5. Hou, C.; Hu, B.; Zhu, J. Photocatalytic degradation of methylene blue over TiO$_2$ pretreated with varying concentrations of NaOH. *Catalysts* **2018**, *8*, 575. [CrossRef]
6. Sareen, D.; Garg, R.; Grover, N. A Study on removal of methylene blue dye from waste water by adsorption technique using fly ash briquette. *IJERT* **2014**, *3*, 610–613.
7. Kumar, R.; El-Shishtawy, R.M.; Barakat, M.A. Synthesis and characterization of Ag-Ag$_2$O/TiO$_2$@ polypyrrole heterojunction for enhanced photocatalytic degradation of methylene blue. *Catalysts* **2016**, *6*, 76. [CrossRef]
8. Rouhi, M.; Babamoradi, M.; Hajizadeh, Z.; Maleki, A.; Maleki, S.T. Design and performance of polypyrrole/halloysite nanotubes/Fe$_3$O$_4$/Ag/Co nanocomposite for photocatalytic degradation of methylene blue under visible light irradiation. *Optik* **2020**, *212*, 164721. [CrossRef]
9. Pan, X.; Chena, X.; Yi, Z. Photocatalytic oxidation of methane over SrCO$_3$ decorated SrTiO$_3$ nanocatalysts via a synergistic effect. *Phys. Chem. Chem. Phys.* **2016**, *18*, 31400–31409. [CrossRef]
10. Gharaei, S.K.; Abbasnejad, M.; Maezono, R. Bandgap reduction of photocatalytic TiO$_2$ nanotube by Cu doping. *Sci. Rep.* **2018**, *8*, 14192. [CrossRef]
11. Theerthagiri, J.; Salla, S.; Senthil, R.A.; Nithyadharseni, P.; Madankumar, A.; Arunachalam, P.; Maiyalagan, T.; Kim, H.S. A review on ZnO nanostructured materials: Energy, environmental and biological applications. *Nanotechnology* **2019**, *30*, 392001. [PubMed]
12. Thermal Energy Radiation Spectrum. Available online: http://agron-www.agron.iastate.edu/courses/Agron541/classes/541/lesson09a/9a.3.html (assessed on 1 September 2020).

13. Karaahmet, O.; Cicek, B. Waste recycling of cathode ray tube glass through industrial production of transparent ceramic frits. *J. Air Waste Manag.* **2019**, *69*, 1258–1266. [CrossRef] [PubMed]
14. Alimohammadi, E.; Sheibani, S.; Ataie, A. Preparation of nano-structured strontium carbonate from Dasht-e kavir celestite ore via mechanochemical method. *J. Ultrafine Grained Nanostruct. Mater.* **2018**, *151*, 147–152.
15. Owusu, G.; Litz, J.E. Water leaching of SrS and precipitation of $SrCO_3$ using carbon dioxide as precipitating agent. *Hydrometallurgy* **2000**, *57*, 23–29. [CrossRef]
16. Lu, P.; Hu, X.; Li, Y.; Zhang, M.; Liu, X.; He, Y.; Dong, F.; Fu, D.; Zhang, Z. One-step preparation of a novel $SrCO_3/g-C_3N_4$ nanocomposite and its application in selective adsorption of crystal violet. *RSC Adv.* **2018**, *8*, 6315–6325. [CrossRef]
17. Divya, A.; Mathavan, T.; Harish, S.; Archana, J.; Milton Franklin Benial, A.; Hayakawa, Y.; Navaneetha, M. Synthesis and characterization of branchlet-like $SrCO_3$ nanorods using triethylamine as a capping agent by wet chemical method. *Appl. Surf. Sci.* **2019**, *487*, 1271–1278. [CrossRef]
18. Suchanek, W.L.; Riman, R.E. Hydrothermal synthesis of advanced ceramic powders. *ASTRJ* **2006**, *45*, 184–193.
19. Khan, S.A.; Shahid, S.; Kanwai, S.; Rizwan, K.; Mahmood, T.; Ayub, K. Synthesis of novel metal complexes of 2-((phenyl(2-(4-sulfophenyl)hydrazono)methyl)diazinyl)benzoic acid formazan dyes: Characterization, antimicrobial and optical properties studies on leather. *J. Mol. Struct.* **2018**, *1175*, 73–89. [CrossRef]
20. Baybars Ali Fil, C.Ö.; Korkmaz, M. Cationic dye (methylene blue) removal from aqueous solution by montmorillonite. *Bull. Korean Chem. Soc.* **2012**, *33*, 3184–3190.
21. Song, L.; Zhang, S.; Chen, B. A novel visible-light-sensitive strontium carbonate photocatalyst with high photocatalytic activity. *Catal. Commun.* **2009**, *10*, 1565–1568. [CrossRef]
22. Moldovan, A.; Neag, E.; Băbălău-Fussa, V.; Cadar, O.; Micle, V.; Roman, C. Optimized removal of methylene blue from aqueous solution using a commercial natural activated plant-based carbon and Taguchi experimental design. *Anal. Lett.* **2018**, *52*, 1–13. [CrossRef]
23. Suqin, L.; Li, W.; Gaopeng, D.; Qiufei, H. Fabrication of $Ag_2CO_3/SrCO_3$ rods with highly efficient visible-light photocatalytic activity. *Rare Metal Mat. Eng.* **2017**, *46*, 0312–0316. [CrossRef]
24. Jin, J.; Chen, S.; Wang, J.; Chen, C.; Peng, T. $SrCO_3$-modified brookite/anatase TiO_2 heterophase junctions with enhanced activity and selectivity of CO_2 photoreduction to CH_4. *Appl. Surf. Sci.* **2019**, *476*, 937–947. [CrossRef]
25. Jin, S.; Dong, G.; Luo, J.; Ma, F.; Wang, C. Improved photocatalytic NO removal activity of $SrTiO_3$ by using $SrCO_3$ as a new co-catalyst. *Appl. Catal. B-Environ.* **2018**, *227*, 24–34. [CrossRef]
26. Castillejos, A.H.E.; De la Cruz Del, F.P.B.; Uribe, A.S. The direct conversion of celestite to strontium carbonate in sodium carbonate aqueous media. *Hydrometallurgy* **1996**, *40*, 207–222. [CrossRef]
27. Zoraga, M.; Kahruman, C. Kinetics of conversion of celestite to strontium carbonate in solutions containing carbonate, bicarbonate and ammonium ions and dissolved ammonia. *J. Serb. Chem. Soc.* **2014**, *79*, 345–359. [CrossRef]
28. Zou, X.; Wang, Y.; Liang, S.; Duan, D. Facile synthesis of ultrafine and high purity spherical strontium carbonate via gas-liquid reaction. *Mater. Res. Express* **2020**, *7*, 025009. [CrossRef]
29. Yang, L.; Chu, D.; Wang, L.; Ge, G.; Sun, H. Facile synthesis of porous flower-like $SrCO_3$ nanostructures by integrating bottom-up and top-down routes. *Mater. Lett.* **2016**, *167*, 4–8. [CrossRef]
30. Wang, Z.; He, G.; Yin, H.; Bai, W.; Ding, D. Evolution of controllable urchin-like $SrCO_3$ with enhanced electrochemical performance via an alternative processing. *Appl. Surf. Sci.* **2017**, *411*, 197–204. [CrossRef]
31. Chen, L.; Jiang, J.; Bao, Z.; Pan, J.; Xu, W.; Zhou, L.; Wu, Z.; Chen, X. Synthesis of barium and strontium carbonate crystals with unusual morphologies using and organic additive. *Russ. J. Phys. Chem.* **2013**, *87*, 2239–2245.
32. Ni, S.; Yang, X.; Li, T. Hydrothermal synthesis and photoluminescence properties of $SrCO_3$. *Mater. Lett.* **2011**, *65*, 766–768.
33. Cao, M.; Wu, X.; He, X.; Hu, C. Microemulsion-mediated solvothermal synthesis of $SrCO_3$ nanostructures. *Langmuir* **2015**, *21*, 6093–6096. [CrossRef]
34. Zhang, J.; Xu, J.; Zhang, H.; Yin, X.; Yang, D.; Qian, J.; Liu, L.; Liu, X. Chemical synthesis of $SrCO_3$ microcrystals via a homogeneous precipitation method. *Micro Nano Lett.* **2011**, *6*, 119–121.
35. Li, L.; Lin, R.; Tong, Z.; Feng, Q. Crystallization control of $SrCO_3$ nanostructure in imidazolium-based temperature ionic liquids. *Mater. Res. Bull.* **2012**, *47*, 3100–3106. [CrossRef]

36. Wang, G.; Wang, L. Different morphologies of strontium carbonate in water/ethylene glycol and their photocatalytic activity. *Fuller. Nanotub. Carbon Nanostructures* **2019**, *27*, 46–51.
37. Alavi, M.A.; Morsali, A. Syntheses and characterization of Sr(OH)$_2$ and SrCO$_3$ nanostructures by ultrasonic method. *Ultrason. Sonochem.* **2010**, *17*, 132–138.
38. Song, L.; Li, Y.; He, P.; Zhang, S.; Wu, X.; Fang, S.; Shan, J.; Sun, D. Synthesis and sonocatalytic property of rod-shape Sr(OH)$_2$8H$_2$O. *Ultrason. Sonochemistry* **2014**, *21*, 1318–1324.
39. Li, S.; Zhang, H.; Xu, J.; Yang, D. Hydrothermal synthesis of f lower-like SrCO$_3$ nanostructures. *Mater. Lett.* **2005**, *59*, 420–422. [CrossRef]
40. Neville, G.A.; Becksteadlf, H.D.; Cooney, J.D. Thermal analyses (TGA and DSC) of some spironolactone solvates. *Fresenius J. Anal. Chem.* **1994**, *349*, 746–750.
41. Stranic, I.; Pang, G.A.; Hanson, R.K.; Golden, D.M.; Bowman, C.T. Shock tube measurements of the tert-Butanol +OH reaction rate and the tert-C$_4$H$_8$OH radical β- scission branching ratio using isotopic labelling. *J. Phys. Chem. A* **2013**, *117*, 4777–4784.
42. Mehrvar, M.; Anderson, W.A.; Young, M.M. Photocatalytic degradation of aqueous organic solvents in the presence of hydroxyl radical scavengers. *Int. J. Photoenergy* **2001**, *3*, 187–191. [CrossRef]
43. Patterson, D.A.; Metcalfe, I.S.; Livingston, F.; Livingston, A.G. Wet air oxidation of linear alkylbenzene sulfonate 2 effect of pH. *Ind. Eng. Chem. Res.* **2001**, *40*, 5517–5525. [CrossRef]
44. Houas, A.; Lachheb, H.; Ksibi, M.; Elaloui, E.; Guillard, C.; Herrmann, J.M. Photocatalytic degradation pathway of methylene blue in water. *Appl. Catal. B Environ.* **2001**, *31*, 145–157. [CrossRef]
45. Rauf, M.A.; Meetani, M.A.; Khaleel, A.; Ahmed, A. Photocatalytic degradation of Methylene Blue using a mixed catalyst and product analysis by LC/MS. *Chem. Eng. J.* **2010**, *157*, 373–378. [CrossRef]
46. Technical Application Guide for PHILIPS LED Lamps. Available online: http://yunyangsh.com/pdf_files/PHILIPS/LED/LED_dengpao/MASTER%20MR16%20LED%2010-50W.pdf (assessed on 31 August 2010).

© 2020 by the authors. Licensee MDPI, Basel, Switzerland. This article is an open access article distributed under the terms and conditions of the Creative Commons Attribution (CC BY) license (http://creativecommons.org/licenses/by/4.0/).

MDPI
St. Alban-Anlage 66
4052 Basel
Switzerland
Tel. +41 61 683 77 34
Fax +41 61 302 89 18
www.mdpi.com

Catalysts Editorial Office
E-mail: catalysts@mdpi.com
www.mdpi.com/journal/catalysts

www.ingramcontent.com/pod-product-compliance
Lightning Source LLC
LaVergne TN
LVHW070629100526
838202LV00012B/759